Paul Wasmer

On the treatment of finite rotations in the discretization of a geometrically exact beam model

Schriftenreihe des Instituts für Mechanik
Karlsruher Institut für Technologie (KIT)

Band 13

Herausgeber:
Prof. Dr.-Ing. habil. Peter Betsch
Prof. Dr.-Ing. habil. Thomas Seelig

Eine Übersicht aller bisher in dieser Schriftenreihe erschienenen Bände finden Sie am Ende des Buches.

On the treatment of finite rotations in the discretization of a geometrically exact beam model

by
Paul Wasmer

KIT Scientific Publishing

Karlsruher Institut für Technologie
Institut für Mechanik

On the treatment of finite rotations in the discretization
of a geometrically exact beam model

Zur Erlangung des akademischen Grades eines Doktor-Ingenieurs
von der KIT-Fakultät für Bauingenieur-, Geo- und Umweltwissenschaften
des Karlsruher Instituts für Technologie (KIT) genehmigte Dissertation

von Paul Wasmer, M.Sc.

Tag der mündlichen Prüfung: 6. Dezember 2024
Hauptreferent: Prof. Dr.-Ing. habil. Peter Betsch
Korreferent: Prof. Dr.-Ing. habil. Sven Klinkel

Impressum

KIT Scientific Publishing

Karlsruher Institut für Technologie (KIT)
KIT Scientific Publishing
Straße am Forum 2
D-76131 Karlsruhe

www.bibliothek.kit.edu/ksp.php | E-Mail: info@ksp.kit.edu | Shop: www.ksp.kit.edu

Print on Demand 2025 – Gedruckt auf FSC-zertifiziertem Papier

ISSN 2363-4936
ISBN 978-3-7315-1410-7
DOI 10.5445/KSP/1000178526

Abstract

The first formulation of the geometrically exact beam was published in 1985 by Simo. Despite its long history, the correct implementation of the model in the Finite Element Method (FEM) and Isogeometric Analysis (IGA) remains a challenging task. The kinematics of the geometrically exact beam require describing the orientation of each of its cross-sections using finite rotations. Finite rotations in three dimensions form the complex mathematical structure of a non-linear manifold. To obtain a path-independent and frame-indifferent model as well as optimal convergence behavior in FEM and IGA, great attention must be paid when discretizing variables lying on a manifold.

This work discusses various discretizations of finite rotations for FEM and IGA, focusing on two possible parametrizations of finite rotations: directors and unit quaternions. Besides a classical, additive discretization, geodesic, and projection-based finite elements are introduced and discussed in detail for both parametrizations. As an additional option for the FEM, elements with shape functions based on Gauss-Lobatto points are introduced.

The numerical behavior of the classical discretization is analyzed alongside a director formulation of the geometrically exact beam using the classical discretization approach for both FEM and IGA. Even though this approach results in a frame-indifferent and path-independent model, the convergence is not optimal for higher-order shape functions for either FEM or IGA. In contrast to this, shape functions based on Gauss-Lobatto points yield the expected convergence order. Additionally, the director formulation is tested on dynamic examples using a structure-conserving time integrator that conserves energy, linear momentum, and angular momentum, yielding excellent results.

In a second application, the geometrically exact beam is formulated within a unit quaternion framework. To achieve optimal convergence behavior, a projection-based approach is employed for both FEM and IGA. Again this formulation is shown to be frame-indifferent and path-independent. The projection-based approach leads to optimal convergence behavior even for higher-order shape functions, unlike the classical discretization approach, which again fails to produce optimal results for higher-order elements. Due to the relatively simple projection-based approach possible with unit quaternions, unit quaternions prove to be an excellent choice in static applications for FEM and IGA.

Kurzfassung

Die erste Formulierung des geometrisch exakten Balkens wurde 1985 von Simo veröffentlicht. Trotz seiner langen Geschichte ist die korrekte Implementierung dieses Modells in der Finite Element Methode (FEM) und der Isogeometrischen Analyse (IGA) immer noch eine Herausforderung. Die Kinematik des geometrisch exakten Balkens erfordert es die Orientierung jedes Querschnittes mittels finiter Rotationen zu beschreiben. Dabei bilden finite Rotationen im dreidimensionalen Raum die komplexe mathematische Struktur einer nichtlinearen Mannigfaltigkeit. Um ein Modell zu erhalten, welches unabhängig vom Bezugssystem und Belastungspfad ist, sowie ein optimales Konvergenzverhalten in FEM und IGA erreicht, muss bei der Diskretisierung von Variablen auf solchen nichtlinearen Mannigfaltigkeiten mit großer Sorgfalt vorgegangen werden.

Diese Arbeit behandelt verschiedene Diskretisierungen finiter Rotationen im Rahmen der FEM und IGA, wobei der Fokus auf zwei mögliche Parametrisierungen finiter Rotationen liegt: Direktoren und Einheitsquaternionen. Neben einer klassischen, additiven Diskretisierung werden 'geodesic' und 'projection-based' finite Elemente eingeführt und für beide Parametrisierungen im Detail diskutiert. Als zusätzliche Möglichkeit für die FEM werden Elemente mit Formfunktionen basierend auf den Gauss-Lobatto Punkten eingeführt.

Das numerische Verhalten der klassischen Diskretisierung wird zusammen mit einer Direktorenformulierung des geometrisch exakten Balkens für FEM als auch IGA betrachtet. Zwar führt diese Formulierung auf ein Modell, welches unabhängig vom Bezugssystem und Belastungspfad ist, jedoch zeigt eine Analyse des Konvergenzverhaltens, dass die klassische Diskretisierung nicht zu einer optimalen Konvergenzordnung bei höheren Ansatzordnungen führt, weder in der FEM noch in der IGA. Im Gegensatz dazu ergeben Formfunktionen basierend auf den Gauss-Lobatto Punkten die erwartete Konvergenzordnung. Zusätzlich wird das Verhalten der Direktorformulierung mit dynamischen Beispielen untersucht, wobei ein strukturerhaltender Integrator verwendet wird, welcher Energie, Impuls und Drehimpuls erhält. Dabei erzielt das Modell hervorragende Ergebnisse.

In einer zweiten Anwendung wird der geometrisch exakte Balken mittels Einheitsquaternionen formuliert. Um ein optimales Konvergenzverhalten zu erreichen, werden hierbei 'projection-based' finite Elemente sowohl für FEM als auch für IGA verwendet. Es wird gezeigt, dass dieser Ansatz wiederum auf eine Formulierung führt, welche System- und Pfadunabhängig ist. Dabei führt der 'projection-based' Elementansatz zu einem optimalen Konvergenzverhalten, auch bei Formfunktionen höherer Ordnung. Im Gegensatz dazu führt der klassische Diskretisierungsansatz, nicht zu optimalen Konvergenzordnungen für Elemente höherer Ordnung. Aufgrund des relativ einfachen 'projection-based' Ansatzes, der mit Einheitsquaternionen möglich ist, erweisen sich Einheitsquaternionen als ausgezeichnete Wahl für Anwendungen der FEM und IGA in der Statik.

Acknowledgments

The presented thesis contains the results of my work as a research associate within the research group of Prof. Peter Betsch at the Institute of Mechanics (IFM) of the Karlsruhe Institute of Technology (KIT). Without the support of many others, this work would not have been possible.

First and foremost, I would like to express my sincere gratitude to my supervisor, Prof. Peter Betsch, for his guidance and confidence in me. His expertise and constructive feedback were instrumental to my work. He provided me with many valuable opportunities to grow both professionally and personally. I am also grateful for the many chances to attend international conferences and summer schools, which allowed me to gain fresh perspectives and broaden my academic horizon.

I would also like to thank Prof. Sven Klinkel (RWTH Aachen) for serving as the second examiner for my thesis and for his genuine interest in my research. Furthermore, I am deeply grateful to Prof. Jose Muñoz (UPC) for welcoming me during a research stay in Barcelona, which was an enriching and inspiring experience.

My thanks go to all my colleagues at the IFM for providing a stimulating and collaborative working environment. Regardless of the topic, there was always a helping hand available. The many discussions were fruitful each time and significantly contributed to the progress of my research. I am particularly grateful for the warm welcome I received when I joined the institute and the many unforgettable events we shared – from attending summer schools and conferences together to engaging in many sports activities and social gatherings.

Special thanks go to Robin Pfefferkorn and Mark Schiebl for proofreading my manuscript and providing valuable feedback.

Finally, and most importantly, I want to thank my family and friends for their unwavering support and unconditional love.

Karlsruhe, January 2025 Paul Wasmer

Contents

1. Introduction

"'Ageometretos medeis eisito': let nobody enter who does not understand geometry. These were the words written at the entrance to Plato's Temple of the Muses. Are numerical analysts welcome in Plato's temple?" [82]

Nowadays 'geometry' has a very broad meaning depending on the field of research and people in ancient Greece most certainly had different concepts in mind when using the word compared to scientists in applied mathematics and mechanics today. Nevertheless, the question remains valid.

1.1. Motivation and methods

Over the last decades, the power of computers has risen exponentially. Due to this increase in computational power, computer simulations no longer need to be performed on large mainframes but can be carried out even on an average laptop. This makes the application of simulation tools much more feasible for many users. Therefore, simulations have become progressively more important in science and engineering, as they allow users to gain insight into the physical behavior of complex systems without the need to plan and perform time-consuming and expensive experiments, while, of course, not replacing them completely. Thus, simulations are a crucial element in the design process of new products but also in research, where they can be used to investigate complex phenomena. Even though the computational power has skyrocketed, it is still important to design simulation methods with the aim of keeping the computational time to a minimum. This can be achieved by either improving the numerical algorithms and their implementation or by using computationally efficient models. Highly efficient simulation methods enable engineers to simulate models in more detail and, thus, perfect their product, while they allow researchers to tackle problems of higher complexity.

One of the most versatile numerical simulation methods is the Finite Element Method (FEM). While the FEM is a well-established method in the industry with a broad range of applications and a solid mathematical foundation, there is still room for improvement. For example, the transition from a design model in Computer Aided Design (CAD) software to a finite element model in finite element analysis software is still a very time-consuming task [32, 78]. A possible approach, which may in the near future help to resolve this issue,

was proposed in 2005 by Hughes *et al.* [78] with the introduction of the Isogeometric Analysis (IGA). The isogeometric concept combines the exact representation of geometries as achieved by CAD software with the versatility and reliability of the FEM in the analysis steps. The IGA is closely related to the FEM as both methods rely on the Galerkin method [32, Chapter 2].
The publication by Hughes *et al.* [78] sparked huge interest in the mechanic and numeric community. Due to the relevance of the topic, a plethora of research concerning the IGA has been conducted and even though a relatively small amount of time passed since its publication, it can be considered a well-established method in the scientific community today. However, a realization in commercial tools is so far not available.
Furthermore, in the course of this research it became evident that in many aspects, the IGA offers many advantages over the FEM [31, 32]. For example, the IGA leads to a higher accuracy per degree of freedom [47]. Nevertheless, it is still important to discuss options to minimize computational time.

Utilizing an appropriate physical model, which reduces the dimension of the problem, can help in this regard. Such models can be obtained through keen physical reasoning, which allows to neglect insignificant physical effects and, thus, reduce the dimension of the problem. For example, one-dimensional models can be used to describe slender three-dimensional geometries. Such one-dimensional models are described by the well-known beam theories.

Beam theories have a long history and date back to the beginning of mechanics at the end of the 17th century, where Jacob I. Bernoulli laid the foundations of the theory [4, Chapter VIII]. Since then most of the beam theories were concerned with only small deformations. Compared to their long history, the history of beam models, which can describe large deformations is relatively short. At the beginning of the 1970's, Reissner [112, 113] made the first advancements to one of the most important beam models for large deformations, the so-called geometrically exact beam. Subsequently, the geometrically exact beam model used today was derived from the three-dimensional theory by Simo in 1985 [126].
Due to the physical restrictions used in the design of the model, it becomes necessary to describe the orientation of the cross-section of the beam with finite or large rotations [126]. Even though the geometrically exact beam model reduces the physical dimension, its mathematical structure is rich as the finite rotations lie on a nonlinear manifold [72]. A basic knowledge of the underlying mathematical structure, which describes finite rotations, is very helpful when dealing with them. The theory that forms their basis is called Lie group theory. The structure of Lie theory does not just cover rotations but many aspects in classical mechanics [72, 73]. Nevertheless, "..., *Lie theory is poorly known in comparison to its importance*" [74]. Even though this quote is already more than 40 years old, the statement still holds, especially for engineers. The knowledge of Lie theory allows us to perform many computations in a linear vector space, instead of the complex, curved geometry of the non-linear manifold [74]. This can be exploited in numerical methods, such as the FEM and IGA.

At the heart of each Galerkin method lies the approximation of the solution with a finite number of basis functions. For models, where the variables lie in a flat linear space the approximation approach is straightforward. However, the complex geometry of finite rotations makes their optimal treatment in Galerkin methods a challenging task. Even though the difficulties accompanied by finite rotations in the FEM were recognized by Simo and Qu-Voc [127, 128] in the very first publication concerning the topic, it took researchers more than a decade to propose the first path-independent and frame-indifferent discretization approach in the context of the geometrically exact beam [33, 83].
Even though subsequent publications discussed the treatment of finite rotations in a finite element framework further (e.g. [11, 21, 51, 62, 114, 117, 135]), only in very recent years the topic was covered by mathematicians for general nonlinear manifolds accompanied by corresponding proofs.
To date, the proposed methods in the literature can be categorized into two major concepts. The first concept is known as geodesic finite elements [120–122]. It is based on the intrinsic distance measure of the underlying manifold on which the approximated field lies. On arbitrary manifolds the geodesic element formulation can only be obtained with the help of implicit algorithms [120].
The second concept relies on a classical, additive approximation approach with a subsequent closest-point or orthogonal projection onto the geometry of the manifold [49, 59]. Elements based on this concept are, therefore, called projection-based elements [59].

Even though both element concepts allow for an arbitrary approximation order, the literature is still very limited, when it comes to higher-order approximation approaches in the FEM to problems on nonlinear manifolds [122]. Accordingly, even less literature exists for approximations of such problems in the context of the IGA. Even though some publications concern with the topic of the beam and IGA, they mostly cover the topic based on collocation methods (e.g. [93, 94, 149]). For the IGA as a Galerkin method, only very little literature regarding the geometrically exact beam formulation exists. Also the discretization approaches for variables describing finite rotations, such as quaternions, are only covered sparsely by the literature. As stated by Tasora [143] "*In fact, the spline interpolation of quaternions is still a debated problem...*". We can generalize this statement by saying the spline interpolation of rotations is, in fact, still a debated problem.

Therefore, in this work, we aim to close this gap and investigate and discuss possible discretization approaches in the FEM and IGA for different types of representations of finite rotations. Subsequently, they are applied in the context of the geometrically exact beam for both approaches. Elements of higher order for both, FEM and IGA, and their behavior are discussed in detail.
For this purpose, we recapture the fundamentals of the IGA and, further, briefly introduce the most important aspects of the Lie group theory, which allows us to discuss finite rotations. Subsequently, we choose two representations of rotational tensors and discuss possible approaches to approximate them in a finite element and isogeometric framework, respectively. Both approaches are applied to the description of the geometrically exact beam. The influence of the chosen discretization for both approaches is investigated in detail.

1.2. Outline

In the beginning of this work, we introduce the isogeometric analysis in detail in **Chapter 2**, where we begin with an elaborated motivation of the topic. Since the IGA relies on B-Splines and NURBS to represent the geometry, these functions are discussed comprehensively. Subsequently, the key aspects of the IGA are discussed, focusing on the topics needed for this work. We cover the possible mesh refinement strategies and present an algorithm from the literature for the computation of the initial configuration. Furthermore, the aspect of the numerical integration in the IGA is discussed. Finally, various advantages of the IGA over the FEM are highlighted.

Chapter 3 reviews the topic of rotations and their discretization. It introduces key aspects of the Lie group theory, which helps to understand the topic in a more general setting. Subsequently, rotations in three dimensions are discussed. Three possible representations of rotations are introduced, where a focus lies on unit quaternions. Afterward, the knowledge of Lie groups and rotations is used to construct finite element discretization approaches for two of the representations introduced beforehand. Four different options for their discretization are discussed: a classical additive discretization, a discretization using the concept of geodesic elements, projection-based elements, and shape functions computed based on the Gauss-Lobatto points as a novel approach.

The model of the geometrically exact beam is introduced in **Chapter 4**, where in the beginning theoretical aspects are discussed alongside a literature survey, which further gives insight into recent developments and employments of the model. Next, the most relevant assumptions of the beam theory are used to derive its kinematic relations in detail. Furthermore, the equations of motion are derived through the balance equations of linear and angular momentum. Subsequently, its kinetic energy and stored-energy function are introduced alongside the employed constitutive. In addition, a brief discussion on the topic of shear constrained Kichhoff-Love beam formulations is presented.

Chapter 5 covers the geometrically exact beam in a director framework. It provides the associated continuous equations, followed by a finite element discretization approach, where two possible options are introduced to enforce the orthonormality constraints needed in this framework. A classical, purely additive discretization is employed for all approximated fields and two options are discussed to enforce the orthonormality constraints are presented. It is proven that this discretization approach leads indeed to a frame-indifferent formulation. Furthermore, a short introduction of the applied structure-conserving time integration scheme is given. The model and its implementation are validated alongside several numerical examples. The convergence behavior for Lagrange and NURBS functions is discussed in detail. Additionally, the results of Gauss-Lobatto-based shape functions are presented. The path-independence of the discretization approach is investigated in a numerical example. Furthermore, Lagrange and NURBS shape functions are used in dynamic examples.

A geometrically exact beam formulation using unit quaternions to parametrize the rotational tensors is introduced in **Chapter 6**. Again the corresponding equations are derived in detail for the continuous case, where it is proven that such an approach indeed leads to frame-indifferent strain measures. A projection-based approach introduced in Chapter 3 is used to discretize the unit quaternions. It is proven that the approach leads to frame-indifferent discretized strain measures. The convergence behavior is investigated in numerical examples for both Lagrange and NURBS shape functions. Again the path-independence is shown alongside a numerical example.

Chapter 7 gives, at last, a summary of the most important findings presented in this work and draws a conclusion. Furthermore, an outlook on possible further research topics is given, where the findings of this work could be applied.

2. Fundamentals of the Isogeometric Analysis

Abstract: At the beginning of this work, the fundamentals of the Isogeometric Analysis (IGA) are recollected. At first, the topic is motivated by briefly discussing the shortcomings of the classical Finite Element Method (FEM) with regard to the mesh generation. Subsequently, B-Splines and NURBS are discussed in Section 2.2 and Section 2.3. In Section 2.4 the IGA is briefly introduced, where the different refinement strategies, the computation of initial control point values as well as the numerical integration are discussed in more detail. At last, the advantages of the Isogeometric Analysis over the classical FEM are discussed.

2.1. Motivation

The physical behavior of continua, such as the deformation of solid bodies, the dispersion of heat, or the propagation of electromagnetic waves, to name just a few, is described by partial differential equations (PDEs). However, only for special cases, a closed-form analytical solution is known. It is thus of utmost importance to have an efficient and reliable solution method at hand to approximate the solutions of PDEs when dealing with many problems in engineering and science. This can be achieved with numerical approximation methods which are indispensable in modern engineering sciences.

A very versatile and reliable numerical method for this task is the FEM. The FEM is a well-established method with an abundance of mathematical proofs verifying the accuracy of the approximated results. It is therefore a widespread tool in industry and science when analyzing physical systems in all fields of engineering and science.

Yet, until today, a bottleneck persists in the overall analysis process, stemming from the challenging and time-consuming transition between CAD and FEM models [32, 78]. The transition often demands a simplification of the CAD model, involving the removal of detailed features, such as small drill holes, grooves, or complex contours, that are assumed to have negligible impact on the overall solution of the problem. This process already demands a resourceful and experienced user. Subsequently, the CAD model has to be transferred to a FEM program, where the geometry is partitioned such that it can be more easily divided into small subdomains, so-called finite elements. This process is known as meshing. Achieving a mesh of reasonable quality is no trivial task. Especially for complicated, curved geometries, deriving a high-quality mesh is challenging, yet necessary for reliable and accurate results. According to Cottrell *et al.* [32, Chapter 1] the preparation of the FEM models takes up to approximately 80% of time of the whole analysis process, while the actual computational time to solve the PDEs only constitutes about 4%. Therefore, it still is a challenging task to obtain simulation results of the current state of design models in a sensible amount of time due to the excessive effort required for model preparation. Moreover, daily changes in the design common to modern design processes increase the issue even further. This limits accurate predictions of the behavior of the real physical system at hand and therefore the relevance of the analysis process.

Hughes *et al.* [78] proposed the concept of the Isogeometric Analysis (IGA) as a possible solution to overcome the mentioned challenges. The FEM and IGA have many things in common since they are both Galerkin methods [32, Chapter 3]. Galerkin methods use a finite set of basis functions to approximate the solutions of differential equations. In the classical FEM, the basis functions are chosen in such a way that the degrees of freedom are directly interpretable and the resulting matrices are sparsely populated. These basis functions are also used to approximate the geometry, which is known as the isoparametric concept. This approach only allows the exact representation of geometries that can be represented by the FEM basis functions. Many formulations of different basis functions exist, however, in general, low-order Lagrange polynomials are used. Regrettably, Lagrange polynomials can precisely represent only a very limited number of geometries.

On the other hand, in the IGA mostly Non-Uniform-Rational-B-Splines (NURBS) are employed. NURBS are usually used as the basis to represent the geometry in CAD programs. Since NURBS are rational functions, they can represent even conic sections exactly. The key idea of the IGA is to use the same NURBS functions representing the geometry as the basis for the approximation of the solution. Therefore, the IGA holds the potential to greatly simplify the transition between a CAD model and the model used for the analysis.

This chapter introduces the fundamentals of NURBS before introducing the concept of the IGA. Computer graphics as well as the IGA are each vast fields of research in themselves; therefore, only a very brief introduction of the most important aspects crucial to the rest of this work can be provided here and we, thus, limit the discussion to one-dimensional domains.

As the name "Non-Uniform-Rational-B-Splines" already suggests, NURBS are based on B-Splines. So before introducing the concept of NURBS, the computation of B-Splines and their various properties as well as B-Spline geometries are discussed in Section 2.2. Using the introduced concept of the B-Splines, a brief introduction of NURBS follows in Section 2.3. Section 2.4 gives more details on the IGA and focuses on important differences to the classical FEM as well as presents additional advantages of the IGA over the FEM.

2.2. B-Splines

Section 2.1 to Section 2.3 closely follow Chapter 2, Section 2.1 and Section 2.2 of Cottrell *et al.* [32]. Further, information is taken from Piegl and Tiller [110, Chapter 1 - 4]. As the given information relies on these two source, they are not cited explicitly.

2.2.1. Knot vector

B-Splines play a key role in NURBS functions and their construction is covered in this section. So at the beginning of this chapter, we start by introducing B-Splines.
Unlike classical finite elements, where the basis functions are constructed on a reference element, B-Splines are constructed using a knot vector. A knot vector is a set

$$\Xi = \{\xi_1, \xi_2, ..., \xi_{n+p+1}\} \tag{2.1}$$

of $n + p + 1$ ordered knot entries $\xi_i \in \mathbb{R}$. Therein, $p \in \mathbb{N}$ is the polynomial order of the basis, and $n \in \mathbb{N}^+$ denotes the number of basis functions. In general, the knot entries can be chosen arbitrarily, however, for convenience, we choose $\xi_1 = 0$ and $\xi_{n+p+1} = 1$, so that $\xi_i \in [0, 1]$, in the remainder of this work.
The entries of the knot vector are sorted in an ascending order such that $\xi_{i-1} \leq \xi_i \leq \xi_{i+1}$. Note that the entries of the knot vector do not have to be unique but can be repeated. A repeated entry in the interior reduces the continuity of the resulting basis locally. This phenomenon will be discussed in more detail in Section 2.2.2.
For geometries of higher dimensions multiple knot vectors, one for each dimension, must be defined. Together, the knot vectors form the parameter space. Many already complex geometries can be represented by using only a single parameter space, to which we refer also as patch. A patch can be thought of as a super element encompassing large parts of the geometry. As geometries become more complex, multiple patches are needed to represent the geometry.
A knot vector is said to be uniform if the knots are spaced equidistant and non-uniform otherwise. The half-open interval $[\xi_i, \xi_{i+1})$ between two knots defines a knot span. Since the knot entries do not need to be unique, this interval may have zero length. In the context of the analysis, we refer to a knot span with a length greater than zero as a finite element. The knot vector, thus, partitions the geometry and the parameter space into finite elements and, thus, defines the mesh.
If the very first ξ_1 and the last ξ_{n+p+1} entry of the knot vector are repeated with a multiplicity of $p + 1$ the knot vector is called an open knot vector. Usually, open knot vectors are employed in combination with the IGA since they are also standard in CAD technology. In the one-dimensional case, the resulting basis of an open knot vector is interpolatory at both ends. This simplifies the connection of various patches as well as the application of boundary conditions for the one-dimensional case.
Even though it is not outright obvious, a knot vector determines many of the properties of the corresponding B-Spline basis. We discuss these properties alongside an example below (see Fig. 2.2).

2.2.2. B-Spline basis

After a knot vector is defined, the associated B-Spline basis functions can be computed using the Cox-de Boor formula. The formula is recursive and must be evaluated multiple times for higher orders. The basis functions of order $p = 0$ are given by

$$N_{i,0}(\xi) = \begin{cases} 1 & \text{for} \quad \xi_i \leq \xi < \xi_{i+1}\,, \\ 0 & \text{otherwise}\,. \end{cases} \tag{2.2}$$

The higher order basis functions of order $p \in \mathbb{N}$ are computed via[1]

$$N_{i,p}(\xi) = \frac{\xi - \xi_i}{\xi_{i+p} - \xi_i} N_{i,p-1}(\xi) + \frac{\xi_{i+p+1} - \xi}{\xi_{i+p+1} - \xi_{i+1}} N_{i+1,p-1}(\xi) \tag{2.3}$$

for $i = 1, ..., n$.

Exemplary B-Spline bases for orders $p = 0, 1, 2$ are shown in Fig. 2.1. An open knot vector $\Xi = \{0, ..., 0.25, 0.5, 0.75, ..., 1\}$ is used as basis for the computation for each order p, where the first and last entry are repeated $p + 1$-times.
The basic properties of the corresponding bases resulting from Eq. (2.2) and Eq. (2.3) are discussed in the following. As can be deduced by examining Eq. (2.2) for order $p = 0$, the basis results in a piecewise constant function, where only one basis function is equal to one in each knot span ($N_{i,0} = 1$ for $[\xi_i, \xi_{i+1})$) while all other basis functions are equal to zero, as displayed in Fig. 2.1a.
When the polynomial order increases, the B-Spline basis becomes a piecewise polynomial, as can be derived from Eq. (2.3). A first-order B-Spline basis results in piecewise linear functions, as shown in Fig. 2.1b. The basis looks very familiar to someone with a basic knowledge of the FEM, and indeed, a first-order B-Spline basis is identical to a first-order discretization approach with classical Lagrange finite elements. Two linear basis functions are nonzero over every knot span, and they are equal to one at each knot, which leads to an interpolatory behavior at the knots.
In Fig. 2.1c the first three basis functions of a second-order B-Spline basis are shown. Examining Eq. (2.3) it is evident that the bases are piecewise polynomials of order p composed of a linear combination of two $p-1$-degree basis functions. Keeping in mind that a knot span might have zero length ($\xi_i = \xi_{i+1}$), it can be seen that the basis functions span multiple knot spans. By inspecting the different basis functions $N_{1,2}$, $N_{2,2}$ and $N_{3,2}$, we can observe that a basis function of order p is greater than zero over $p + 1$ knot spans. This is due to the so-called local support property of the B-Spline basis functions, which can be expressed by

$$N_{i,p}(\xi) \begin{cases} \neq 0 & \text{for} \quad [\xi_i, \xi_{i+p+1})\,, \\ = 0 & \text{otherwise}\,. \end{cases} \tag{2.4}$$

[1] where the following is defined $\frac{(\bullet)}{0} := 0$

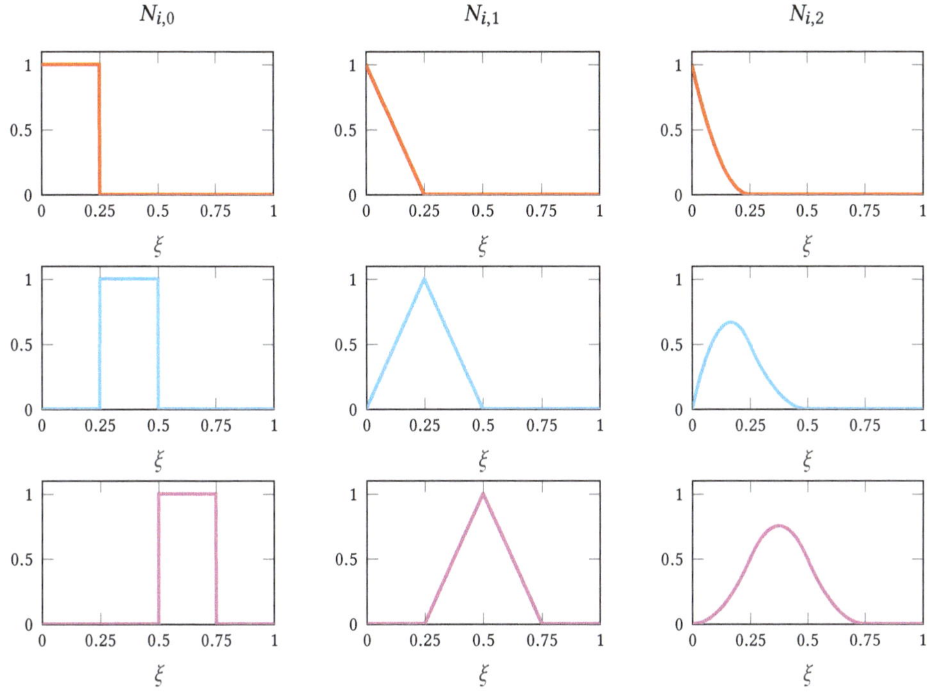

(a) B-Spline basis functions of order $p = 0$

(b) B-Spline basis functions of order $p = 1$

(c) B-Spline basis functions of order $p = 2$

Figure 2.1.: Results for an exemplary B-Spline basis of different orders for the first three basis functions. The associated open knot vector is given by $\Xi = \{0, \ldots, 0.25, 0.5, 0.75, \ldots, 1\}$, where the first and last entry are repeated $p + 1$-times.

From the local support property, we can derive that on each knot span a maximum of $p + 1$ shape functions $N_{i,p}$ are nonzero. This property can be exploited to design efficient algorithms, meaning that not all basis functions need to be evaluated at every point ξ of the knot vector. Pseudocode can be found in the literature e.g. [110, Chapter 2].

Fig. 2.1 further shows that none of the plotted functions is less than zero. Having a closer look at Eq. (2.2) and Eq. (2.3) we see that this indeed holds for all basis functions. The nonnegativity of each shape function

$$N_i(\xi) \geq 0 \qquad \forall\, \xi \in \left[\xi_1, \xi_{n+p+1}\right] \tag{2.5}$$

is another important property of the B-Spline basis functions.

Futhermore, the B-Spline basis fulfills the partition of unity

$$\sum_{i=1}^{n} N_{i,p}(\xi) = 1 \qquad \forall\, \xi \in \left[\xi_1, \xi_{n+p+1}\right] . \tag{2.6}$$

This is a fundamental requirement, when combining the B-Spline basis with the Finite Element Method as it allows to correctly represent rigid body movements and constant strains in classical continuum bodies [77].
Furthermore, we can see that the plotted B-Spline basis functions for $p > 0$ have only one global maximum. This is indeed true for all B-Spline basis functions of order $p \geq 1$.

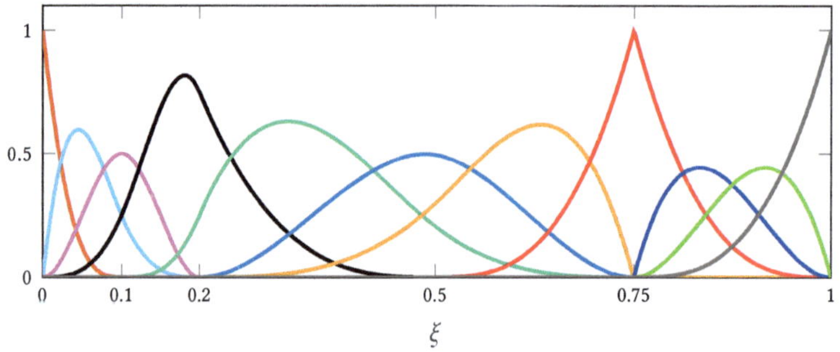

Figure 2.2.: Example of a B-Spline basis of order $p = 3$ in the parameter space, with the knot vector $\Xi = \{0, 0, 0, 0, 0.1, 0.2, 0.2, 0.5, 0.75, 0.75, 0.75, 1, 1, 1, 1\}$.

We want to discuss some of the properties mentioned above and the influence of a higher multiplicity of knot vectors on a specific example. For this purpose a knot vector is chosen as $\Xi = \{0, 0, 0, 0, 0.1, 0.2, 0.2, 0.5, 0.75, 0.75, 0.75, 1, 1, 1, 1\}$ for which the resulting B-Spline basis of order $p = 3$ is shown in Fig. 2.2.
The chosen knot vector is an open knot vector, as can be seen from the multiplicity of $p + 1$ of the first and last entries. This results in the interpolatory behavior of the first basis function at $\xi = 0$ and respectively of the last basis function at $\xi = 1$. Furthermore, there exist two more knot values with a multiplicity greater than one. $\xi_6 = \xi_7 = 0.2$ with a multiplicity $m = 2$, and $\xi_9 = \xi_{10} = \xi_{11} = 0.75$ with a multiplicity $m = 3$. The influence of the triple entry 0.75 is directly apparent. It reduces the continuity of the basis to C^0 at $\xi = 0.75$, which leads to a single function being equal to one at this point. The influence of the double entry at $\xi = 0.2$ is not as easily visible. The continuity of the basis is here reduced to C^1. This reduction becomes obvious in the derivative displayed in Fig. 2.3. We can deduce that the continuity of the basis is reduced by each knot multiplicity by one due to Eq. (2.2). In a formal manner, if the multiplicity of a knot entry is given by m, the continuity over a knot is given by C^{p-m}, which is another noteworthy property shown of the B-Spline basis. It allows the construction of curves with interpolatory behavior as shown in Section 2.2.4.
Furthermore, we can see the local support property of the B-Splines in Fig. 2.2 since every basis function N_i is only nonzero in $[\xi_i, \xi_{i+4})$. As mentioned above each basis function is equal to or greater than zero and has exactly one global maximum.

2.2.3. Derivatives of B-Splines

Besides the basis, the derivates of it are needed for various applications in a CAD tool, as well as for the application in an analysis step. The first derivative of the basis functions is given by

$$\frac{\partial}{\partial \xi} N_{i,p}(\xi) = \frac{p}{\xi_{i+p} - \xi_i} N_{i,p-1}(\xi) - \frac{p}{\xi_{i+p+1} - \xi_{i+1}} N_{i+1,p-1}(\xi) \,. \tag{2.7}$$

Further differentiating Eq. (2.7) leads to an equation for the k−th derivative of $N_{i,p}(\xi)$, which can be calculated by

$$\begin{aligned} \frac{\partial^k}{\partial \xi^k} N_{i,p} &= \frac{p}{\xi_{i+p} - \xi_i} \frac{\partial^k}{\partial \xi^k} N_{i,p-1} - \frac{p}{\xi_{i+p+1} - \xi_{i+1}} \frac{\partial^k}{\partial \xi^k} N_{i+1,p-1} \\ &= \frac{p!}{(p-k)!} \sum_{j=0}^{k} a_{k,j} N_{i+j,p-k} \end{aligned} \tag{2.8}$$

with [2]

$$a_{0,0} = 1 \,, \tag{2.9}$$

$$a_{k,0} = \frac{a_{k-1,0}}{\xi_{i+p-k+1} - \xi_i} \,, \tag{2.10}$$

$$a_{k,j} = \frac{a_{k-1,j} - a_{k-1,j-1}}{\xi_{i+p+j-k+1} - \xi_{i+j}} \quad \text{for} \quad j = 1, \ldots, k-1 \,, \tag{2.11}$$

$$a_{k,k} = \frac{-a_{k-1,k-1}}{\xi_{i+p+1} - \xi_{i+k}} \,. \tag{2.12}$$

In Fig. 2.3 the first derivative of the exemplary basis from Fig. 2.2 is shown. In contrast to the basis, the derivative can take on negative values. Due to the partition of unity of the basis, the sum of all derivatives is equal to zero.
As before, the influence of the multiplicity of the knot vector entries is visible. At $\xi = 0.2$, where a double entry exists, the derivative of the basis is C^0-continues and at $\xi = 0.75$ a discontinuity of N_7 to N_9 occurs.
Equivalent to the algorithm for the Cox-de Boor formula, efficient algorithms for the evaluation of the derivatives exist in the literature exploiting the local support property (see e.g. [110, Chapter 2]).

[2] where we define $\frac{(\bullet)}{0} := 0$

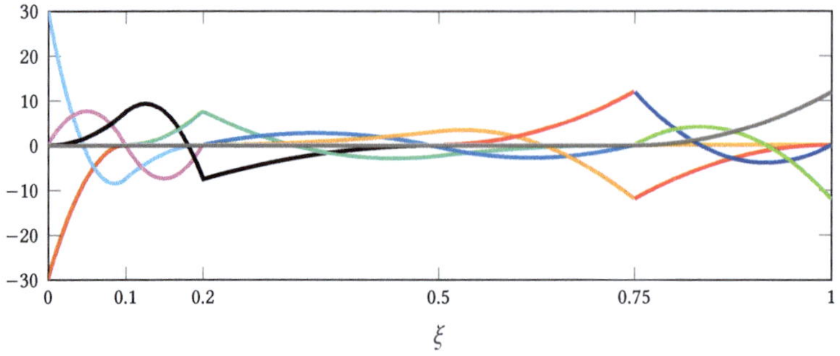

Figure 2.3.: First derivate of the exemplary B-Spline basis of order $p = 3$ from Fig. 2.2 in the parameter space with the knot vector $\Xi = \{0, 0, 0, 0, 0.1, 0.2, 0.2, 0.5, 0.75, 0.75, 0.75, 1, 1, 1, 1\}$

2.2.4. B-Spline geometries

A B-Spline basis can be used to represent many geometric objects by introducing so-called control points $\mathbf{B}_i \in \mathbb{R}^d$, where $2 \leq d \in \mathbb{N}$ is the dimension in which the geometry is represented. The number of control points is equivalent to the number of basis functions. The B-Spline curve is represented via

$$\mathbf{C}(\xi) = \sum_{i=1}^{n} N_{i,p}(\xi)\mathbf{B}_i \,. \tag{2.13}$$

The curve $\mathbf{C}(\xi)$ does not interpolate the control points $\mathbf{B}_i$ for arbitrary basis orders $p \geq 2$.

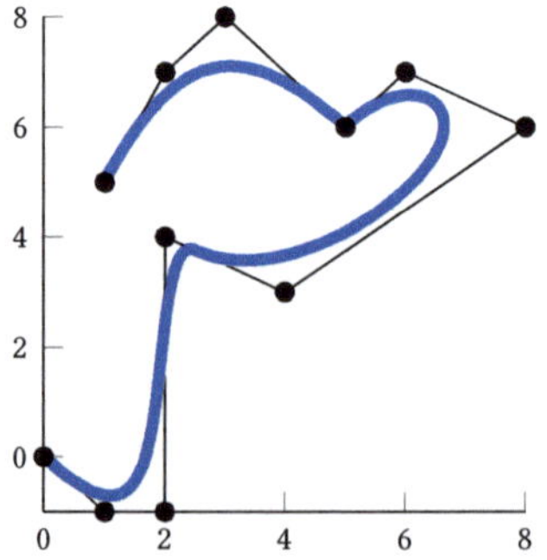

Figure 2.4.: Example of a B-Spline curve in two dimensions with the basis from Fig 2.2

We discuss the properties of the resulting B-Spline curve alongside an example (see Fig. 2.4). The same B-Spline basis as in Fig. 2.2 is used. The corresponding eleven control points are collected in the columns of the following matrix

$$\mathbf{B} = \begin{bmatrix} 0 & 1 & 2 & 2 & 4 & 8 & 6 & 5 & 3 & 2 & 1 \\ 0 & -1 & -1 & 4 & 3 & 6 & 7 & 6 & 8 & 7 & 5 \end{bmatrix} .$$

In Fig. 2.4 the control points are connected with a black line. The resulting interpolation is called the control polygon, which forms a linear approximation of the B-Spline curve. The B-Spline curve in Fig. 2.4 lies inside the convex hull of the control points. This is a general property of B-Spline curves resulting from the properties of the basis.
Since the basis is computed with an open knot vector, the first and last control points are interpolated, whereas all interior control points except the eighth control point ($\mathbf{B}_8 = \begin{bmatrix} 5 & 6 \end{bmatrix}^\mathsf{T}$) are not. The interpolary nature of the curve at the eighth control point results from the chosen knot vector, where the knot $\xi = 0.75$ has a multiplicity of $m = 3$. It is directly related to the fact that at $\xi = 0.75$ only the basis function N_8 is non-zero while all other functions are zero (see Fig. 2.2). Thus, $\mathbf{C}(\xi)$ is interpolatory at $\mathbf{B}_8$.

Geometries of higher dimensions are constructed by applying the tensor product. Surfaces described with B-Splines are thus given by

$$\mathrm{S}(\xi, \eta) = \sum_{i=1}^{n} \sum_{j=1}^{m} N_{i,p}(\xi) M_{j,q}(\eta) \mathbf{B}_{i,j} \tag{2.14}$$

and volumes by

$$\mathrm{S}(\xi, \eta, \zeta) = \sum_{i=1}^{n} \sum_{j=1}^{m} \sum_{k=1}^{l} N_{i,p}(\xi) M_{j,q}(\eta) L_{k,r}(\zeta) \mathbf{B}_{i,j,k} , \tag{2.15}$$

where $M_{j,q}$ and $L_{k,r}(\zeta)$ are the j-th and k-th basis function of a basis of order q and r respectively. A geometry defined with only a single set of basis functions over one knot vector for each dimension can already describe complex geometries. However, the complexity of the geometries is limited. More complex geometries can be represented by using multiple patches and combining the resulting geometries.

Affine transformations, such as rotations, translations, scalings, or, shears, of a B-Spline curve are performed by applying the affine transformation to the control net. This is an essential property used in the proof of objectivity in the IGA as well as for the fulfillment of the patch-test [78]. It is known as affine covariance.
Algorithms for the computation of the coordinates of the control points as well as many tools for the manipulation of the curves can be found in the literature, e.g. [110].

2.3. Non-uniform rational B-Splines

As mentioned above, B-Splines are still piecewise polynomials and even though many geometries can be represented with them, they cannot represent all geometries exactly. Conic sections, for example, cannot be exactly described using polynomials. However, rational functions can fill this gap. Thus, the concept of B-Splines was expanded towards so-called Non-Uniform Rational B-Splines (NURBS).

The basis for NURBS are B-Splines. Additionally so-called weights $w_i \in \mathbb{R}^+$ are introduced for the definition of a NURBS basis [3]. The i-th NURBS basis function of order p is computed from

$$R_{i,p}(\xi) = \frac{N_{i,p}(\xi)w_i}{W(\xi)} = \frac{N_{i,p}(\xi)w_i}{\sum_{\hat{i}=1}^{n} N_{\hat{i},p}(\xi)w_{\hat{i}}} \tag{2.16}$$

with $i = 1, ..., n$ and where $N_{i,p}$ is the i-th B-Spline basis functions of order p as introduced in Section 2.2.2. If all weights are set equal to one ($w_i = 1 \, \forall \, i$), the NURBS basis reduces to the corresponding B-Spline basis. Due to the computation from a B-Spline basis the properties illustrated for B-Splines in Section 2.2.2 carry over to the NURBS basis. Therefore, they are not repeated here again.

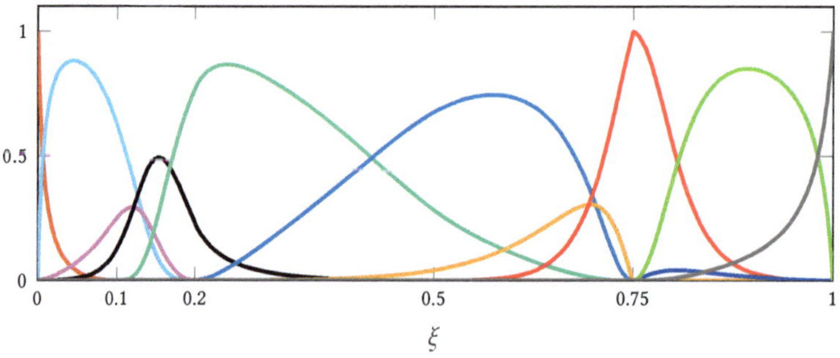

Figure 2.5.: Example of a NURBS basis in the parameter space.

Fig. 2.5 shows a NURBS basis based on the B-Spline basis from Fig. 2.2. The following arbitrary weights have been chosen for the NURBS basis

$$\mathbf{w} = \begin{bmatrix} 1 & 5 & 1 & 1 & 10 & 10 & 1 & 1 & 0.1 & 4 & 1 \end{bmatrix}.$$

When comparing Fig. 2.2 with Fig. 2.5, it can be seen, that the NURBS basis is a modified version of the B-Spline basis. However, all the relevant properties mentioned in Section 2.2 still hold.

Equivalent to a B-Spline curve, a NURBS curve $\mathbf{C}(\xi)$ is described by

$$\mathbf{C}(\xi) = \sum_{i=1}^{n} R_{i,p}(\xi)\mathbf{B}_\mathbf{i} \, . \tag{2.17}$$

In the following, we briefly discuss the influence of the weights on the resulting geometry. Again we demonstrate this on an example. The influence of a change of the weights is illustrated in Fig. 2.6, where the same basis as in Fig. 2.5 is employed and only weight w_5

[3] $w_i > 0$ is an often made assumption, but other choices are possible [110].

changes. The same control points as in Fig. 2.4 are used. The original B-Spline curve from this example is plotted for comparison in a dashed light blue line.
A large value of a weight corresponds to a stronger influence of the associated control point. An increase pulls the NUBRS curve towards the corresponding control point. If on the other hand, the weight is decreased the influence of the control point on the overall geometry is reduced since the curve does not run as closely by it.
The weight of the fifth control point w_5 is varied for each NURBS curve in Fig 2.6. For better visibility, the position of the associated control point $\mathbf{B}_5$ is shown in red. In Fig. 2.6a the weight is set to $w_5 = 0.01$, thus the influence of the fifth control point on the curve is low. The curve runs very close by the fourth and sixth control points while running in a nearly straight line between both. When the weight is increased to one ($w_5 = 1$) the influence of the control point increases, and the curve is drawn a little closer to the control point as can be seen in Fig. 2.6b. Due to the relatively high weight of w_6, it does not run as closely by as the original B-Spline curve. For a weight $w_5 = 10$ the curve bends very closely towards the corresponding control point thus increasing its influence on the surrounding geometry, which is shown in Fig. 2.6c. It shows, that the influence of the fourth control point $\mathbf{B}_4$ is reduced alongside the increase of the influence of control point $\mathbf{B}_5$. Similar effects can be examined for $\mathbf{B}_6$.

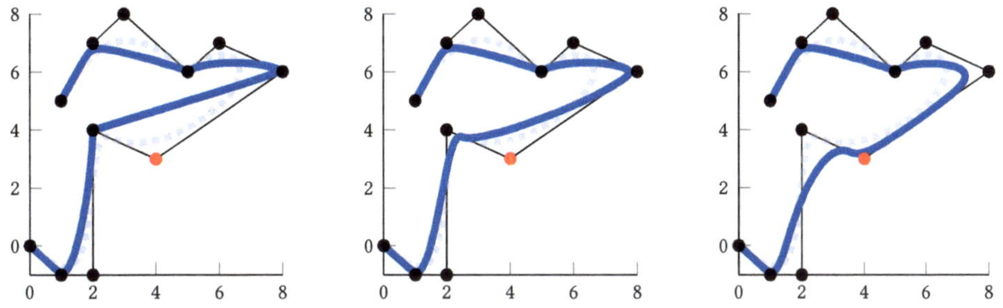

(a) NURBS curve with $w_5 = 0.01$ **(b)** NURBS curve with $w_5 = 1$ **(c)** NURBS spaces curve with $w_5 = 10$

Figure 2.6.: NURBS curve with variation of the fifth weight w_5. The fifth control point $\mathbf{B}_5$ is highlighted in red. The original B-Spline curve from Fig. 2.4 is plotted for comparison dashed in light blue.

Surfaces and volumes can be constructed in the same manner, by applying the tensor product as shown for B-Splines in Section 2.2.4. As affine transformations are applied to the control net, the affine covariance holds the same as for B-Spline curves.

2.4. The Isogeometric Analysis

Using the concepts from the last two sections, we can now turn our attention again to the Isogeometric Analysis. Like the classical FEM, the IGA is a method to approximate the solution of PDEs. The underlying mathematical approach is equivalent to the FEM

because both are Galerkin methods. Galerkin methods rely on the approximation of the solution space with a finite set of functions. The classical FEM approach uses approximation functions, which are only locally nonzero. Shape functions, which fulfill the collocation properties on the finite element nodes are chosen for the approximation, which enables a conforming discretization over element boundaries. The most common approach uses Lagrange polynomials, which are constructed using the nodal values on a reference element. Lagrange polynomials result in a C^0-continues discretization over the element boundaries, however, there are methods to achieve higher continuity in one dimension (e.g. Hermite shape functions [77]).
In the classical FEM, the basis functions are used to describe the geometry via the isoparametric concept. So the FEM basis functions are used to approximate the solution and are also used to represent the geometry. It follows directly that only geometries, which can be described through the piecewise continuous basis functions, are represented exactly.

In the case of the IGA, this idea is reversed. Instead of using the approximation functions from the Galerkin approach for the representation of the geometry, the functions used to display the geometry are also used to approximate the solution space. In general, these are B-Splines, NURBS, or some of their further advancements. As introduced above they are not defined using elements and nodes but through knot vectors in the parameter space.
Both, the FEM and the IGA, are Galerkin methods as mentioned above. Thus, the approximation in general is given by an additive approach[4]

$$\mathbf{u}_{\text{exact}} \approx \mathbf{u}^h = \sum_{i=1}^{n} N_i \mathbf{u}_i , \tag{2.18}$$

where $\mathbf{u}_{\text{exact}}$ is the exact solution, $\mathbf{u}^h$ is the approximated field, and $\mathbf{u}_i$ are the values of the weights corresponding to the approximation function N_i. In the case of the FEM, $\mathbf{u}_i$ corresponds to the nodal value at node i, and for the IGA $\mathbf{u}_i$ corresponds to the value at the control point i. N_i are either FEM basis functions, like Lagrange polynomials, or in the case of IGA B-Spline or NURBS functions. For convenience, we denote a shape function from here on always with N_i for either Lagrange, B-Spline, or NURBS functions.
However, in contrast to a FEM basis function, which interpolates the nodes, a B-Spline or NURBS curve does not interpolate the control points. It follows that the degrees of freedom $\mathbf{u}_i$, which lie on the control points, cannot be interpreted as the approximated solution. Instead the discretization approach of Eq. (2.18) has to be employed. This is in contrast to FEM, where due to the interpolatory nature of the basis functions the nodal values $\mathbf{u}_i$ coincide with the approximated solution at the nodes.

In Fig. 2.7 the most important spaces needed for the IGA are shown. Besides the physical space, which is described with the isoparametric concept through a linear combination of the basis function with the control points, the parameter space and a reference element are displayed. As explained in Section 2.2, the bases are constructed in the parameter space,

[4] In Chapter 3 we are going to introduce discretization approaches, which do not rely purely on an additive approach.

which is defined through knot vectors. The knot vectors also separate the parameter space, as well as the physical space into finite elements, through interior knot entries as mentioned in Section 2.2.2. One of the finite elements is highlighted in blue in Fig. 2.7 in the parameters space as well as in the physical space. The element is defined through the knot spans $[\xi_{p+1} = 0, \xi_{p+2} = 0.5)$ and $[\eta_{p+1} = 0, \eta_{p+2} = 0.25)$ with $p = 3$. This is in contrast to the classical finite element basis functions, where finite elements are obtained by subdividing the geometry in the physical domain. Furthermore, the B-Spline or NURBS basis is constructed in the parameter space, whereas classical finite elements are usually constructed on a reference element. Even though the basis is constructed in the parameter space, still a reference element is needed. It is used for the numerical integration process. A discussion on the numerical integration in the IGA follows in Section 2.4.3.

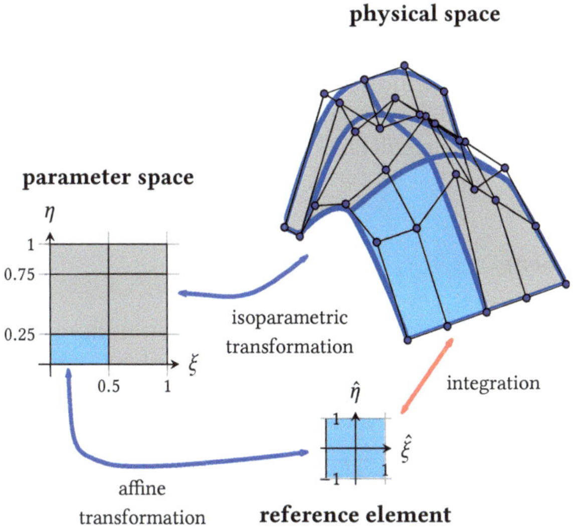

Figure 2.7.: Spaces for the application of the IGA.

Due to the very similar structure of the mathematical approach, only small adjustments to an existing FEM code are necessary to adapt it to the IGA. Besides a loop over the finite elements, an additional loop over different patches, which form complex geometries is needed. Furthermore, the shape function on each element can differ, which leads to the necessity to select different functions for each element. Details on the issues of implementation are given in [32, Appendix A].

Even though the basis functions are now defined on a patch level and not on the local level of finite elements the local support property of B-Spline and NURBS ensures that the bandwidth of the resulting matrices still does not increase compared to the FEM [32, Chapter 2]. However, numerical experiments indicate that the computational time increases nevertheless [109].

Like the FEM the IGA has a solid foundation through mathematical proofs, which show the convergence of the approximated result towards the exact solution with mesh refinement (see e.g. [13, 35]).

2.4.1. Refinement strategies

Mesh refinement strategies in the context of the IGA deserve a closer look. The IGA offers more possibilities compared to the classical FEM. This allows for an overall greater flexibility of the method. We use here the definitions of the refinement strategies as introduced by Hughes [78].
Besides the classical h-refinement, the refinement in space, and the classical p-refinement, where the order of the shape functions is increased while keeping the overall continuity of the shape functions over the boundaries constant, the IGA offers a third option. This third option is named k-refinement.
The k-refinement increases the order of the shape functions together with the overall continuity of the discretization over the whole patch. In the following, we are going to discuss the three options in more detail.

2.4.1.1. h-refinement

Equivalent to the FEM, the possibility to choose a finer mesh reducing the size of the finite elements exists. For this purpose, additional unique knots are inserted into the knot vector without changing the basis order.

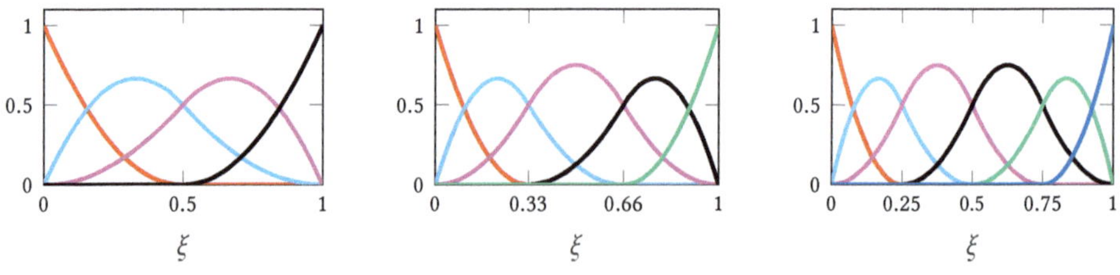

(a) Two finite elements **(b)** Three finite elements **(c)** Four finite elements

Figure 2.8.: An example of the h-refinement strategy. The B-Spline basis of order $p = 2$ is C^1-continues over the whole domain. The continuity is kept constant, while a smaller length of the elements (or knot spans) leads to an increase in the number of finite elements.

The h-refinement is illustrated in Fig. 2.8. In Fig. 2.8a a second-order B-Spline basis with two finite elements is shown. We now subdivide the domain into three elements, where we chose equidistant knots. This leads to Fig. 2.8b with three finite elements. If we further divide the domain into four elements, we obtain the basis displayed in Fig. 2.8c. The order

of the basis is kept constant and so is the continuity inside each element (C^2) and over element boundaries (C^1) since only unique knots are introduced. With each additional knot, the number of basis functions increases by one and consequently, the number of control points increases accordingly.
In the FEM additional elements can be added by subdividing the geometry into smaller pieces. In the FEM the meshing process in one-dimensional problems is straightforward. In contrast to this, the h-refinement in the IGA is already difficult for one-dimensional domains, if the exact geometrical representation is to be preserved. It makes the use of algorithms necessary to compute the position of the newly introduced control points and weights. Corresponding algorithms can be found e.g. in [78, 110].

2.4.1.2. p-refinement

In the standard FEM, p-refinement refers to an increase in the polynomial order of the shape function on each element. Due to the element-wise construction process, the increase in the polynomial order does not lead to an increase in the continuity of the basis over element boundaries. Using classical approaches, the continuity between elements remains C^0. This the definition of p-refinement is employed by Hughes [78].

The described behavior can be mimicked by the IGA by introducing additional knot entries inside the interior of a knot span when increasing the element order. To achieve this the multiplicity of the interior knots is increased alongside the order of the basis functions to keep the continuity constant.
An example of the p-refinement strategy is shown in Fig. 2.9. In Fig. 2.9a, a B-Spline basis of order $p = 2$ is plotted. At $\xi = 0.5$ it is C^0-continues due to a knot entry of multiplicity of $m = 2$ ($\xi_4 = \xi_5 = 0.5$). An increase of the basis order to $p = 3$ is shown in Fig. 2.10b. Additionally, a knot is introduced, increasing the knot multiplicity to $m = 3$ ($\xi_5 = \xi_6 = \xi_7 = 0.5$). A further p-refinement as defined in [78] results in Fig. 2.10c, where a basis of order $p = 4$ is drawn. To ensure the same continuity over element boundaries the multiplicity is increased to $m = 4$ ($\xi_6 = \xi_7 = \xi_8 = \xi_9 = 0.5$).

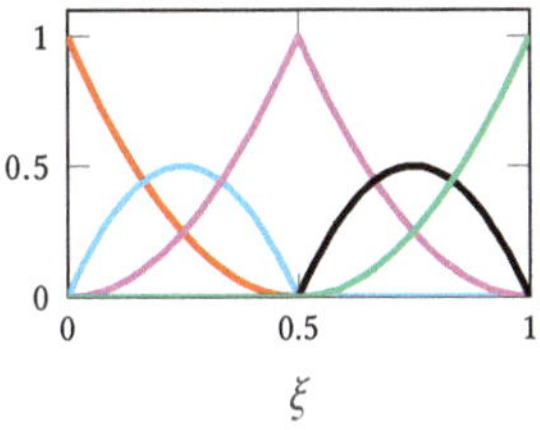

(a) C^0-continues B-Spline basis of order $p = 2$

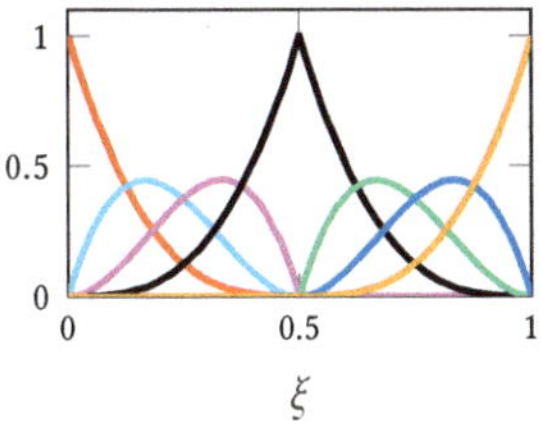

(b) C^0-continues B-Spline basis of order $p = 3$

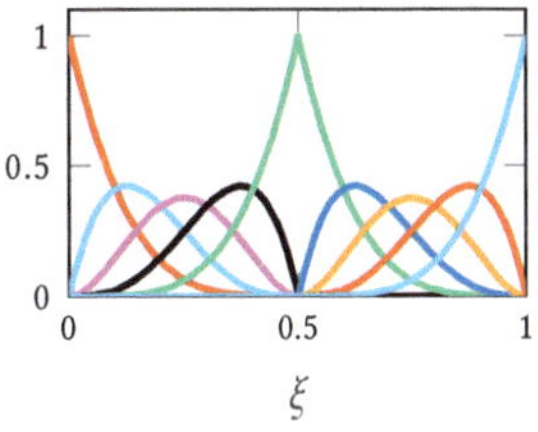

(c) C^0-continues B-Spline basis of order $p = 4$

Figure 2.9.: An example of the p-refinement strategy of a B-Spline basis. The basis is C^0-continues over element boundaries. The continuity is kept constant, while the order of basis is increased.

2.4.1.3. *k*-refinement

While h- and p-refinement strategies exist in a similar fashion in FEM and IGA, the later offers an additional possibility for another new refinement approach. Hughes *et al.* [78] named this new refinement strategy k-refinement. A direct comparison of the effects of k-refinement versus p-refinement can be found for example in [31, 80, 111].

Instead of just increasing the order on each element individually, k-refinement increases the continuity of the basis functions over the whole patch. While, in general, this is not possible in the FEM, in the IGA it can be done by elevating the basis order. In contrast to the h- or p-refinement no additional interior knots are introduced. A classical finite element formulation has no equivalent counterpart.

An illustration of the k-refinement strategy is shown in Fig. 2.10. We begin with a second-order B-Spline basis, as shown in Fig. 2.10a, which is C^1-continues over the element boundary at the single knot entry at $\xi_4 = 0.5$. By increasing the order of the basis the overall continuity is increased. Fig. 2.10b displays an increase of the basis to order three. Since no further internal knots are introduced the basis over the element border at $\xi = 0.5$ also increases to C^2. A further increase of the order leads to Fig. 2.10c, where the order is $p = 4$. Thus, at $\xi = 0.5$ a C^3-continuity is given.

We discuss some advantages of this approach below in Section 2.4.4. Higher order finite elements with B-Spline or NURBS basis functions in this work are constructed using k-refinement.

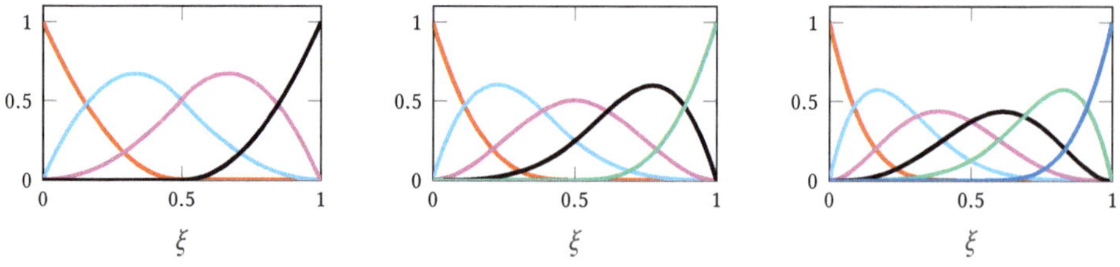

(a) C^1-continues B-Spline basis of order $p = 2$

(b) C^2-continues B-Spline basis of order $p = 3$

(c) C^3-continues B-Spline basis of order $p = 4$

Figure 2.10.: An example for the k-refinement strategy of a B-Spline basis over two elements. Continuity over element boundaries and basis order are increased simultaneously.

2.4.2. Computation of initial control point values

In the IGA the control points, which correspond to the degrees of freedom, generally may not correspond to the approximated solution directly. Due to the non-interpolatory nature, the approximated solution is given through the linear combination of the values on the control points multiplied by the shape functions (Eq. (2.18)). This has to be taken into account in the computation of the initial configuration in a preprocessing step. The position of the control points and the corresponding weights to represent complex geometry exactly can be obtained through the use of CAD software. However, simple geometries such as circles, conics, or, cylinders can be constructed with the aid of simple algorithms given e.g. in [110].

In addition to the control point values required for the positions, the values for the other fields, e.g., initial rotational parameters, the initial velocity distribution, or initial temperature, have to be computed in a separate step. An algorithm for the computation of the initial director values in the context of shells at the control points is proposed in [41]. It can be generalized from directors to arbitrary fields and is briefly repeated here. We assume that all relevant fields are known in the initial configuration, such that the values at the Gauss points $\mathbf{u}^h(\xi_{\mathrm{GP}})$ can be determined in every element. A value of an arbitrary approximated field $\mathbf{u}^h$ inside a one-dimensional domain is approximated at the Gauss points by

$$\mathbf{u}^h(\xi_{\mathrm{GP}}) = \sum_{i=1}^{n} N_i(\xi_{\mathrm{GP}})\mathbf{u}_i\,, \tag{2.19}$$

where $\mathbf{u}^h(\xi_{\mathrm{GP}})$ are the approximated values, evaluated at the Gauss point level, and ξ_{GP} are the position of the Gauss points at the level of the parameter space. The unknowns $\mathbf{u}_i$ are the values on the control points. Moreover, N_i are either B-Splines or NURBS functions. Eq. (2.19) can be written in matrix notation as

$$\mathbf{u}^{\mathrm{GP}} = \mathbf{N} \cdot \mathbf{u}^{\mathrm{CP}}\,, \tag{2.20}$$

where the matrix $\mathbf{N}$ has the dimension $(n_{\mathrm{GP}} \cdot (n_{el} \times n_{\mathrm{CP}}))$. The number of Gauss points per element is given by n_{GP} and n_{el} is the number of finite elements. The vector $\mathbf{u}^{\mathrm{GP}}$ contains all the values on the Gauss points and $\mathbf{u}^{\mathrm{CP}}$ all values on the control points, respectively. By multiplying Eq. (2.20) from the left with $\mathbf{N}^\top$ we obtain the following equation

$$\mathbf{N}^\top \cdot \mathbf{u}^{\mathrm{GP}} = \mathbf{N}^\top \cdot \mathbf{N} \cdot \mathbf{u}^{\mathrm{CP}}\,. \tag{2.21}$$

The values of the approximated field $\mathbf{u}$ on the control points are obtained by an inversion of the form

$$\mathbf{u}^{\mathrm{CP}} = (\mathbf{N}^\top \cdot \mathbf{N})^{-1} \cdot \mathbf{N}^\top \cdot \mathbf{u}^{\mathrm{GP}}\,. \tag{2.22}$$

where the matrices $\mathbf{N}$ and $\mathbf{N}^\top \cdot \mathbf{N}$ can be assembled elementwise. Note that a sufficiently high number of Gauss points needs to be chosen, since otherwise, the system might be ill-conditioned [41].

The presented algorithm will be used in Chapter 5 and Chapter 6 to compute the initial field of directors and quaternions for an application in the geometrically exact beam formulation.

2.4.3. Numerical integration

In every finite element routine, it is necessary to evaluate integrals. For special cases, the integration can be performed analytically, but, in general, numerical approaches are needed. The standard algorithm for the numerical integration is the Gaussian quadrature. It is an optimal integration scheme for polynomials in one-dimensional problems since it uses a minimal number of integration points to achieve an exact solution [77].
However, in the case of the FEM or IGA, this optimality only holds for a singular element. Even for a finite element discretization approach with interpolatory shape functions fewer integration points are needed for an exact integration [79]. Thus, the globally optimal integration scheme for a finite element patch has to be defined over element boundaries. Due to the higher continuity of B-Splines and NURBS, the efficiency gain of new integration rules could be even higher than for the classical C^0-continues elements [79]. This fact, in itself, calls for the search for new, more efficient quadrature rules. Furthermore, the employed functions in the IGA are, in general, rational functions as shown in Section 2.3. This implies, that even on an element level in one dimension the classical Gauss integration is not optimal since it is not exact, but only gives an approximated value.

This sparks the search for new, more efficient integration schemes for splines, as well as for the exact quadrature rules of rational functions [70]. An exact quadrature rule for rational functions can be achieved through an expansion of the concept of the Gaussian quadrature. Such an expansion is called a generalized Gaussian quadrature, which is again a quadrature rule, that integrates functions, exactly, while no other rule exists, that integrates those functions with fewer integration points. It is shown in [85] that such an integration rule exists for a large class of functions including rational functions like NURBS.
However, the more efficient quadrature rules existing in the literature have some severe drawbacks. In contrast to quadrature rules on reference elements, which can be applied to arbitrary meshes, the corresponding weights and integration points have to be computed for each mesh anew. Further, they often have very strict requirements on the underlying mesh in order to compute the necessary tools for integration accurately see e.g. [7, 9, 84]. The integration points and weights of these quadrature rules must be computed numerically. The non-linear systems, which need to be solved, are numerically ill-posed and, thus, hard to solve [70, 79]. The mentioned shortcomings greatly limit the applicability of the newly developed rules.

Due to the limited success in finding a generalized Gaussian quadrature rule for arbitrary meshes, to the best knowledge of the author, the Gaussian quadrature is still employed in the IGA as standard approach. Nevertheless, the excellent results of the IGA in every field, show that the error of the numerical integration can be neglected. Furthermore, the Gaussian quadrature is not always the optimal rule for higher dimensional Lagrange finite elements [77]. Also, the standard FEM approach may not integrate polynomials depending on the underlying problem. This might for example occur, for distorted meshes, where the Jacobian is rational. Likewise, non-polynomial functions need to be evaluated when complex non-linear material behavior like elastic-plastic material is investigated [14].

For the reasons mentioned above, the classical Gauss integration scheme is applied in this work for the integration of NURBS and B-Spline functions alike.

2.4.4. Advantages of the Isogeometric Analysis

The main motivation for the IGA especially from the viewpoint of the industry, is the optimization of the workflow (see Section 2.1). Due to different representations in CAD programs, where solids are represented through their surface, and FE analysis tools, where a volumetric representation is needed, this is still an open topic of research. Besides the described optimization of the workflow, the IGA offers other advantages, that might be of greater interest to the scientific community.

The IGA could not only enable a faster transition between CAD model and analysis but also allows an exact representation of the geometry. This is especially of interest for the solution of non-linear differential equations. It is well-known, that the solutions for non-linear problems may strongly vary for a slight variation of the initial conditions. It directly follows that the solution can highly depend on the representation of the initial geometry [3].

2.4.4.1. Higher continuity of shape functions

As mentioned before, the IGA also allows for additional possibilities in the design process of the basis functions. In the classical Finite Element Method, the construction of bases is performed element-wise. It is, therefore, a very challenging task to design a basis with high continuity over elements. On the other hand, NURBS shape functions are formulated in a more global context. Their construction process allows for a straightforward computation of a basis with arbitrary continuity on the level of patches. This is especially of interest in higher dimensions, where it is not possible to construct such a basis in the classical FEM.

As presented in the previous section, the construction of a discretization, which is smooth over element boundaries is naturally embedded in the IGA in the form of k-refinement. This allows for a broader application of the method compared to the standard FEM. Weak forms of PDEs, which result in a variational index greater than one, require a discretization of a continuity higher than C^0. Differential equations, which lead to such a requirement, are for example the Cahn-Hilliard equation [32, 69], equations describing materials with gradient elasticity [48] or the Navier-Stokes-Korteweg equation [52]. But more familiar for engineers are problems from structural analysis such as shear stiff beams [50, 99] and shells [42, 86] with the requirement of a C^1-continues basis.
Apart from such applications, contact mechanics is a field where the application of the IGA is promising. Problems involving contact still remain a challenging task. Besides many other difficulties, one issue is a discretization with low smoothness over element boundaries [37]. The low smoothness may lead to a sudden switch of the direction of

the contact forces, leading to difficulties in the convergence of the non-linear solution [37]. The advantages of a smooth discretization are, for example, shown in [103] for the simulation of contact with B-Splines.

2.4.4.2. Error bounds

Another advantage of the IGA is its overall lower error bounds compared to the FEM, which is shown numerically on various examples in e.g. [47]. The IGA in combination with k-refinement leads to a higher accuracy per degree of freedom when compared to the classical FEM. We show the higher accuracy with data taken from an example, where the results are presented in Section 6.5.1 [5]. The convergence plot of FEM with Lagrange elements and IGA with NURBS are presented in Fig. 2.11. The example is solved for orders $p = 1, 2, 3$ for both methods. For linear elements $p = 1$ both methods coincide, thus the lead to the same result. An increase in the element order leads to different approximations. While the order of convergence is the same for both methods, an overall lower error bound can be found for the IGA with the same number of degrees of freedom. It can be concluded that the IGA is more efficient per degree of freedom.

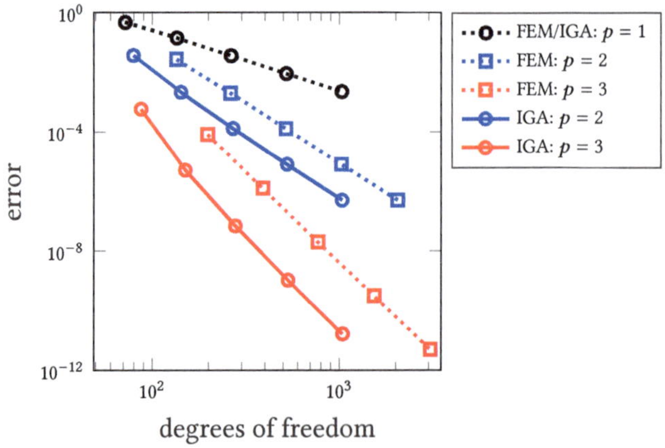

Figure 2.11.: Exemplary convergence behavior of FEM vs IGA for approximation order of $p = 1, 2, 3$. The FEM approach is discretized with Lagrange shape function, whereas for the IGA approximation B-Splines are used. For $p = 1$ both approaches coincide.

[5] Results presented here are obtained with the NPS(•) and LPS(•) elements introduced in Section 6.5.

2.4.4.3. Simulation of wave propagation

In addition to the already mentioned advantages, the IGA is beneficial, when dealing with dynamic problems. The basis functions used in the IGA do not exhibit the so-called Runge's phenomenon. The Runge's phenomenon describes the behavior found for Lagrange basis functions of higher order, where shape functions show strong oscillations, especially, towards the element boundaries. This might lead to ill-conditioned matrices depending on the discretization approach [133].
In contrast to this even B-Spline bases of very high order show a smooth behavior [32, Chapter 2]. Thus, basis functions of arbitrary order can be chosen. This is of great advantage when dealing with dynamic problems, where higher order elements can approximate wave propagation phenomena much more efficiently than lower orders [150].

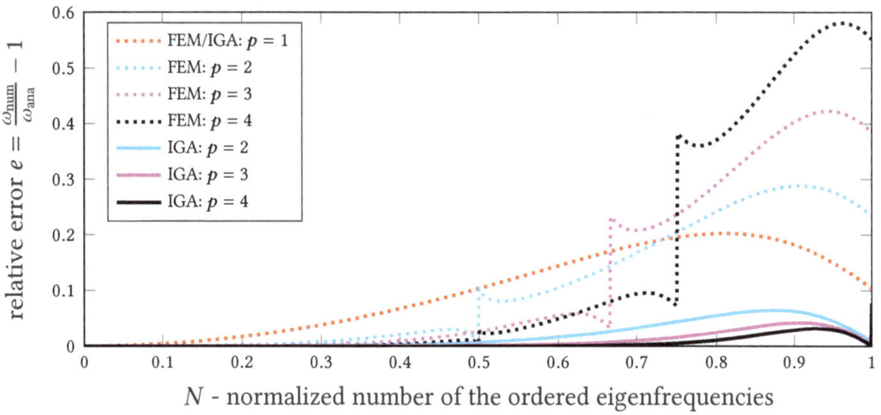

Figure 2.12.: Relative error of the approximated eigenfrequencies of a linear rod over the normalized number of degrees of freedom. The FEM computation is performed using C^0-continues Lagrange shape functions, while in the IGA computation C^{p-1}-continues B-Spline shape functions are used.

The higher continuity of B-Splines and NURBS has another advantage. Dynamic simulations with finite elements using Lagrange shape functions show an unfavorable behavior. The eigenfrequencies show a switch between so-called acoustic and optical modes [6]. This can be seen when investigating the approximated eigenfrequencies. For this purpose, we reproduce results found in [31, 32]. The oscillatory behavior of a linear elastic rod fixed at both ends is investigated. In Fig. 2.12 the relative error $e = \omega_{\text{num}}/\omega_{\text{ana}} - 1$ of the numerical eigenfrequencies ω_{num} is plotted over the normalized number of eigenfrequencies N. ω_{ana}

[6] The terms acoustic/optical modes are borrowed from crystallography, where these modes describe the oscillation of crystals of sodium and chlorine ions, which resonate in very different frequencies. The acoustic frequencies are of lower frequency near the audible range, whereas the optical frequencies oscillate with very high frequencies [32, Chapter 5].

are the eigenfrequencies of the analytical solution. The total number of degrees of freedom of the systems is held constant at 999 for all element orders.
Both methods coincide for $p = 1$. The resulting curve displays an increase of the error up to approximately $N = 0.8$, with an already significant error at around $N = 0.4$. Nevertheless, the curve is smooth over the whole spectrum.
Higher-order elements lead to a much better approximation for $N < 0.5$. The FEM solution results in slightly larger errors than the IGA. At approximately $N = 0.5$, however, the FEM solutions exhibit a sudden increase in frequency. Higher order approximations lead to an additional jump at around $N = 0.65$ for $p = 3$ and at approximately $N = 0.75$ for $p = 4$. These solutions can be regarded as spurious. In a spline-based approximation, such a jump does not occur. The solution displays an overall smooth behavior with a much smaller error over the whole spectrum.
The jump found in the FEM solution can be explained due to a switch from acoustic to optical modes. While the acoustic modes are modes, where the nodes oscillate in phase over element boundaries, the optical branch comes from out-of-phase oscillations of the internal element nodes. Since only the boundary nodes of each element are in contact, this can lead to localized vibration patterns. In a spline-based approach, the influence of each node spans multiple elements. Thus, the behavior of the individual elements is closely connected to its neighbors, which circumvents the localized patterns. An exception from this is the highest frequency at $N = 1$, where a jump in the relative error occurs. This can be explained by the usage of an open knot vector, where the most outer control points can exhibit a localized vibration pattern as well.

2.4.5. Conclusion

Overall the IGA has many benefits over the classical FEM. In general, it enables more choices for the analysis. However, this comes at a cost. The IGA approach is much more complex, especially when it comes to the initialization and refinement of the mesh. The mesh refinement is especially complicated in higher dimensions (see e.g. [69]). However, many of those challenges might be addressed, when the IGA becomes incorporated into commercial tools.

3. Rotations in three dimensions

Abstract: In this chapter, we give a short introduction to the topic of rotations and their treatment in the FEM and IGA. For this purpose, a very brief summary of the most relevant aspects of the Lie group theory needed for this work is given. Further, the special orthogonal group is discussed in detail as well as some of its parametrizations. The focus lies hereby on unit quaternions. Finally, the treatment of rotations in the context of the FEM and IGA is examined, where four discretization approaches are discussed. The first approach is a classical, additive discretization, the second approach is based on the concept of geodesic finite elements and the third uses projection-based finite elements. Additionally, we briefly present Gauss-Lobatto-based shape functions as an option for the FEM.

At first glance, rotations[1] might seem like a trivial subject since we deal with them in daily life, where we instinctively know how to handle them. However, their treatment in a mathematical framework is complex and involves many aspects of advanced mathematics. Rotations play a key role in many engineering tasks as they are needed to describe the movement of rigid bodies. Consequently, they are one of the key topics in the field of multibody dynamics (cf. [10]). Furthermore, rotational fields are needed in infinite dimensional problems. Examples range from structural mechanics, where rotations are used to describe the deformation behavior of beams and shells [41, 102], to some special cases of continuum problems, such as Cosserat continua [146].

Even though it is a key feature of many engineering tasks, the topic is in general not covered in great detail in the education of engineers. In this chapter, we thus want to introduce the basics of rotations in three dimensions. For this purpose, we want to give the reader a very brief insight into the underlying mathematical concept of the Lie group theory. A basic knowledge of this concept, even though it is often very abstract and not trivial, simplifies the treatment of rotations, as it unites various aspects of the topic.

Furthermore, we give an insight into different approaches, which can be used for the

[1] In the literature, it is often emphasized, that large or finite rotations are treated in comparison to small or infinitesimal rotations. Infinitesimal rotations lie in a vector space. Thus, the topic is much simpler in itself. This work focuses on large or finite rotations. The amendment "large" or "finite" is thus dropped for convenience, and if rotations are mentioned they are assumed to be large or finite if not explicitly stated otherwise.

treatment of rotational fields in the framework of the FEM and IGA. To the best knowledge of the author, a discussion for a general treatment of rotational fields in the Isogometric Analysis is so far not given in the literature. This gives us the mathematical foundations we need in Chapter 5 and Chapter 6, where we have a detailed look at the geometrically exact beam formulation and its discretization in a finite element framework for two different formulations of the geometrically exact beam.

This chapter is structured as follows. At first, the fundamental concept of Lie groups is introduced in Section 3.1. These basics are further used to have a more detailed look at rotations in three dimensions in Section 3.2, where subsequently, three possible ways to represent rotations are briefly discussed. At last, we approach the discretization with the FEM and IGA for two of these representations in Section 3.4, where we discuss the classical, additive discretization, geodesic, and projection-based finite elements. Additionally, as an option for the FEM, we introduce Gauss-Lobatto-based shape functions. Their use has so far not been discussed in the literature in connection with this topic.

3.1. Lie groups

The origin of the concept of Lie groups dates back to the work of the Norwegian mathematician Sophius Lie (1842-1899) [82]. They form a very useful but abstract concept when working with rotations and motions. The underlying mathematics, however, goes far beyond the curriculum of mathematical classes of engineers, and even though there exists numerous literature (e.g. [61, 71, 74, 89, 141]) on the topic, it is usually not addressed at engineers or even applied mathematicians. This makes the transfer of the knowledge to numerical applications difficult.
Motivation to engage in the Lie group theory can be found in the literature, where the theory is applied to enhance numerical methods. For example in [26, 28, 82, 96] the concept of Lie groups is applied for the development of time integration schemes and allows the conservation of the underlying geometry. An application to variational integrators in combination with the geometrically exact beam formulation is given in e.g. [39, 91], where highly efficient methods are obtained. Sonenville *et al.* [134–136] use the knowledge of Lie groups to develop a locking-free finite element formulation of the geometrically exact beam.

It is far beyond the scope of the present work to give an in-depth and extensive introduction to the topic, but it is merely the goal to give the reader a first insight into the concept. The focus hereby lies on the aspects, which are useful to understand rotations and their treatment in the FEM and IGA, which are subsequently required to solve the problem of the geometrically exact beam in the context of these methods. At the beginning of this section, the definition of Lie groups is given, while the rest of the section gives a short introduction to the most important aspect of the theory needed in this work.

We begin with the definition of a Lie group: A Lie group is defined as a smooth manifold, with a group structure, where the group action and the group inversion are smooth [141, Chapter 1].
We describe each aspect in detail in this section and begin with a descriptive explanation of a manifold. For a mathematical precise definition, see e.g. [95, Chapter 4]. A manifold $\mathcal{M}$ of dimension m can be viewed as an abstract, smooth hypersurface. Locally, at point $\mathbf{x} \in \mathcal{M}$ a manifold is flat and forms a linear vectors space $\mathbb{R}^m$, which we call the tangent space $T_{\mathbf{x}}\mathcal{M}$ at $\mathbf{x}$. The admissible derivatives and variations lie in the corresponding tangent space $T_{\mathbf{x}}\mathcal{M}$. In this work, we focus on embedded manifolds, which are defined through a constraint function. For more details see e.g. [72, Chapter 2].
To help the reader understand each concept presented in this section different examples are given. A very simple Lie group and, thus, a useful example for this purpose, is the unit sphere S^1 [141, Chapter 1], which will be used on each new concept as an example throughout this section.

Example 3.1.1 (Unit sphere S^1). *A Lie group, which every engineer knows well, even though not as Lie group, is the unit sphere S^1 (see Fig. 3.1). The unit sphere S^1 is a circle with a radius equal to one in the complex plane. A parametrization of the unit sphere is given by the complex numbers of unit length*

$$S^1 := \{r \in \mathbb{C} \mid \|r\| = 1\},$$

where $\|r\| = 1$ is the constraint defining the group. The geometry of the unit sphere is curved, however, locally it resembles a flat space. This is the tangent space $T_{\mathrm{r}}S^1$ at r. The sphere itself is described through a one-dimensional curve in a two-dimensional embedding space.

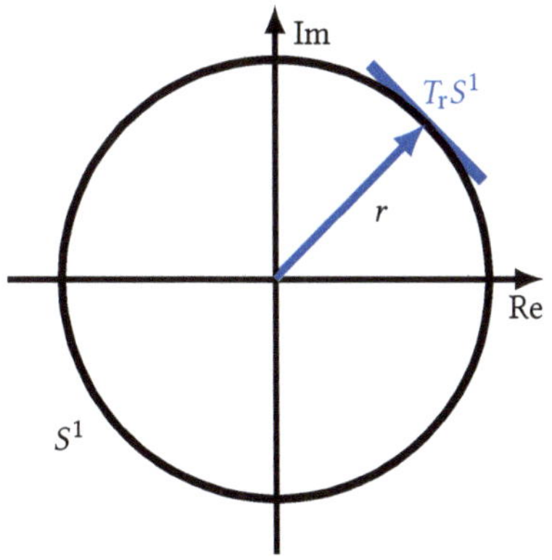

Figure 3.1.: Sketch of the unit sphere S^1 in the complex plane.

Example 3.1.2 (Unit sphere S^2). *Another example of a manifold, which can easily be visualized, is the unit sphere S^2 in $\mathbb{R}^3$ as shown in Fig. 3.2. It forms a smooth surface in the $\mathbb{R}^3$, the embedding space, defined by the constraint $\|\mathbf{x}\| = 1$ with $\mathbf{x} \in \mathbb{R}^3$*

$$S^2 := \{\mathbf{x} \in \mathbb{R}^3 \mid \|\mathbf{x}\| = 1\}.$$

We can associate at any point $\mathbf{x}$ *of* S^2 *a tangent space* $T_{\mathbf{x}}M$*, which is flat (cf. Fig. 3.2). In contrast to* S^1*, however, the unit sphere* S^2 *has no group structure and, thus, is of course no Lie group.*
The two-dimensional geometry can be described using e.g. two angles coordinates. Unfortunately, such a description with two parameters always has singularities [102]. As stated by the Hairy Ball theorem [43, 101] there exists no continuous non-vanishing tangential vector field on the sphere so it is not smooth. This complicates working on the sphere S^2 *as can be seen in the context of the Kirchhoff-Love beam [50, 118] or Reissner-Mindlin shells [102].*

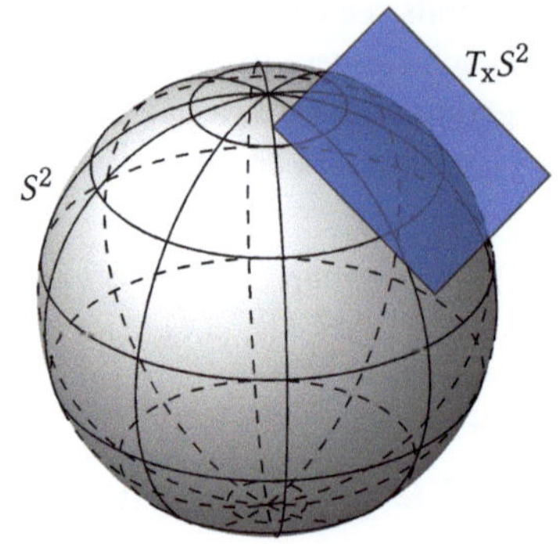

Figure 3.2.: Sketch of the unit sphere S^2.

3.1.1. Group structure

As given in the definition, each Lie group forms a group with its associated action. A group $(\mathcal{G}, \circ)$ is a set of elements $(\mathbf{X}, \mathbf{Y}, \mathbf{Z} \in \mathcal{G})$ in combination with an action $(\circ)$, which fulfill the following properties [72, Chapter 2]:

1. The group is closed under its group action $(\circ)$: $\mathbf{X} \circ \mathbf{Y} \in \mathcal{G}$.
2. There exist an identity element $\epsilon \in \mathcal{G}$, such that $\epsilon \circ \mathbf{X} = \mathbf{X} \circ \epsilon = \mathbf{X}$.
3. Every element $\mathbf{X}$ of $\mathcal{G}$ has an inverse in $\mathbf{X}^{-1} \in \mathcal{G}$ satisfying $\mathbf{X} \circ \mathbf{X}^{-1} = \mathbf{X}^{-1} \circ \mathbf{X} = \epsilon$.
4. The group's action is associative. Thus, $(\mathbf{X} \circ \mathbf{Y}) \circ \mathbf{Z} = \mathbf{X} \circ (\mathbf{Y} \circ \mathbf{Z})$ holds for all elements $\mathbf{X}, \mathbf{Y}, \mathbf{Z} \in \mathcal{G}$.

The concept of groups can be further split into subcategories. One such subcategory are abelian groups, where a group $\mathcal{G}$ is called an abelian or commutative group if $\mathbf{X} \circ \mathbf{Y} = \mathbf{Y} \circ \mathbf{X}$ holds for all $\mathbf{X}, \mathbf{Y} \in \mathcal{G}$ [72]. Abelian groups have a very familiar structure since most mathematics in school fall into this category. The already familiar group of the unit sphere S^1 is an abelian group.

Example 3.1.3 (Unit sphere S^1). *The unit sphere S^1 has a group structure. Its group action is the multiplication of complex numbers denoted by (∗), which makes it an abelian group since $r * z = z * r$ for $r, c \in \mathbb{C}$. The associated identity element is given by $\epsilon = 1 + i0$, and the inverse of a complex number $r = a + ib \in \mathbb{C}$ of unit length is given by its complex conjugate $r^{-1} = \bar{r} = a - ib$.*

Another very familiar abelian group is the group of real numbers $\mathbb{R}$.

Example 3.1.4 (Real numbers $\mathbb{R}$). *A simple example of a group is the group of real numbers $\mathbb{R}$, where the group action is given by the addition (+). For every $X, Y \in \mathbb{R}$ the group operation is closed: $X + Y \in \mathbb{R}$. The identity element ϵ is 0 as $X + 0 = 0 + X = X$ for every $X \in \mathbb{R}$. The inverse of every $X \in \mathbb{R}$ is given by its negative value $X^{-1} = -X$ so that $X + X^{-1} = X - X = 0$. At last, the addition is associative $(X + Y) + Z = X + (Y + Z)$ holds for all $X, Y, Z \in \mathbb{R}$. Since, furthermore, the addition is commutative $X + Y = Y + X$ the real numbers are an abelian group under addition.*

Another Lie group, which is of interest to the topic at hand, is the general linear group GL [72, Chapter 2]. Tensors describing rotations are a subset of this group.

Example 3.1.5 (General linear group $GL(n, \mathbb{R})$). *The general linear group $GL(n, \mathbb{R})$ consists of invertible, matrices in $\mathbb{R}^{n \times n}$ under the operation of matrix multiplication $(\cdot)$. The group's identity is the $n \times n$ identity matrix and the inverse of a group element $\mathbf{A} \in GL(n, \mathbb{R})$ is given by the matrix inverse $\mathbf{A}^{-1} \in GL(n, \mathbb{R})$. The general linear group, however, is a non-abelian group since the matrix multiplication is not commutative $\mathbf{A} \cdot \mathbf{B} \neq \mathbf{B} \cdot \mathbf{A}$ for all $\mathbf{A}, \mathbf{B}, \in GL(n, \mathbb{R})$.*

3.1.2. Lie algebra

Since a Lie group is a smooth manifold, we know we can define a tangent space at every point of the group. Of special interest is the tangent space $T_\epsilon \mathcal{G}$ at the Lie groups identity element ϵ. Even though, Lie groups are, in general, curved, the flat tangent space $T_\epsilon \mathcal{G}$ almost describes its corresponding Lie group combined with the exponential and logarithmic map completely [95, Chapter 9]. These maps will be introduced in the next section.

The tangent space $T_\epsilon \mathcal{G}$ is called Lie algebra of the Lie group $\mathcal{G}$ and is denoted by $\mathfrak{g}$ [95, Chapter 9]. The structure of Lie algebra is revealed when taking the derivative with respect to time of the constraint function defining the Lie group [2]. For the multiplicative Lie groups treated in this work, the derivative with respect to time of the group constraint yields a new constraint of the form

$$\dot{\mathbf{X}}^{-1} \circ \mathbf{X} + \mathbf{X}^{-1} \circ \dot{\mathbf{X}} = \mathbf{0}, \tag{3.1}$$

[2] We omit a definition using the commutator.

where we denote the partial differential with respect to time by $\frac{\partial(\bullet)}{\partial t} = (\dot{\bullet})$. When evaluating the constraint in Eq. (3.1) at the identity $\mathbf{X} = \epsilon$, we obtain

$$\dot{\mathbf{X}}^{-1} + \dot{\mathbf{X}} = \mathbf{0}\,, \tag{3.2}$$

which reveals the structure of the Lie algebra. Objects, which fulfill Eq. (3.2), lie on the Lie algebra $\mathfrak{g}$. We denote the elements of the Lie algebra of group $\mathcal{G}$ with the tilde operator, i.e. $\tilde{\mathbf{v}} \in \mathfrak{g}$.
Again the idea of the Lie algebra is demonstrated using the unit sphere S^1.

Example 3.1.6 (Unit sphere S^1). *The unit sphere S^1 can be parametrized with a complex number $r \in \mathbb{C}$ of unit length $\|r\| = \bar{r} * r = 1$. By differentiating the group constraint with respect to time one obtains*

$$\dot{\bar{r}} * r + \bar{r} * \dot{r} = 0\,.$$

Evaluating the previous equation at the identity element by setting $r = \epsilon = 1$ reveals the structure of the Lie algebra $\mathfrak{s}^1$

$$\dot{\bar{r}} + \dot{r} = 0\,.$$

Therefore, the derivative $\dot{r}$ is equal to its negative conjugate. Thus, $\dot{r}$ lies in the space of purely imaginary numbers, and the Lie algebra is given by (see Fig. 3.1)

$$T_\epsilon S^1 = \mathfrak{s}^1 := \{z \in \mathbb{C} \,|\, \mathrm{Re}(z) = 0\}\,. \tag{3.3}$$

The Lie algebra $\mathfrak{s}^1$ is isomorph to the linear space of real numbers $\mathbb{R}$. Computation on the Lie algebra can, thus, be performed in the space of real numbers.

For all matrix Lie groups, which are the ones relevant for mechanics, the Lie algebra is a useful concept, since there exists a vector space isomorph to it [74]. The corresponding proof for finite-dimensional Lie algebras is given by Ado's theorem [89]. In combination with the exponential and logarithmic map, it gives us the necessary utensils to deal with Lie groups in numerical methods. A quote from Howe [74] underlines the importance of the concept: "*Since $\mathcal{G}$ is a complicated nonlinear object and $\mathfrak{g}$ is just a vector space, it is usually vastly simpler to work with $\mathfrak{g}$. Otherwise intractable computations may become straightforward linear algebra. This is one source of power of Lie theory.*"
More examples of Lie algebras follow in Section 3.2 and Section 3.3.3.

3.1.3. Exponential and logarithmic map

Lie groups are especially convenient since we know the mapping from the flat Lie algebra back onto the curved geometry of the Lie group. We can derive it by evaluating the following differential equation

$$\dot{\mathbf{X}}(t) = \mathbf{X}(t) \circ \tilde{\mathbf{v}}\,, \tag{3.4}$$

where $\mathbf{X} \in \mathcal{G}$ and, where $\tilde{\mathbf{v}} \in \mathfrak{g}$ is taken to be constant [132]. Therein, $\mathcal{G}$ is a Lie group and $\mathfrak{g}$ its associated Lie algebra. Eq. (3.4) is an ordinary differential equation, which can be solved using an exponential ansatz [74, 132]. The solution of Eq. (3.4) is, thus, given by

$$\mathbf{X}(t) = \mathbf{X}(0) \circ \exp_{\mathcal{G}}(\tilde{\mathbf{v}}(t)) . \tag{3.5}$$

By taking into account the group structure, one can deduce that $\exp_{\mathcal{G}}(\tilde{\mathbf{v}}(t))$ must be an element of the group $\mathcal{G}$, since $\mathbf{X}(t)$ and $\mathbf{X}(0)$ are group elements. We call $\exp_{\mathcal{G}} : \mathfrak{g} \mapsto \mathcal{G}$ the exponential map of the corresponding Lie group $\mathcal{G}$. In the case of matrix Lie groups, which we use in this work, the exponential map can be written as [61]

$$\exp_{\mathcal{G}}(\mathrm{v}) = \sum_{j=1}^{\infty} \frac{\tilde{\mathbf{v}}^j}{j!} .$$

This is a very useful concept since a group's manifold is nearly fully described by the group's Lie algebra and the exponential map [95, Chapter 9]. It becomes even more useful since a map from the manifold onto the Lie algebra is also known. It is the inverse of the exponential map known as the logarithmic map $\log_{\mathcal{G}} : \mathcal{G} \mapsto \mathfrak{g}$

$$\log_{\mathcal{G}}(\mathbf{X}) = \tilde{\mathbf{v}} , \tag{3.6}$$

where $\mathbf{X} \in \mathcal{G}$ and $\tilde{\mathbf{v}} \in \mathfrak{g}$. For a matrix Lie group, the logarithmic map can be written as [61]

$$\log_{\mathcal{G}}(\mathbf{X}) = \sum_{j=1}^{\infty} (-1)^{j+1} \frac{(\mathbf{X} - \mathbf{I})^j}{j} . \tag{3.7}$$

The exponential and logarithmic map can be written explicitly in a closed form for some groups by using a Taylor expansion. This is not in general possible for arbitrary matrix Lie groups of higher dimensions [82].
Again the concept is demonstrated for the unit sphere S^1.

Example 3.1.7 (Unit sphere S^1). *The exponential map of the unit sphere S^1 is given by the exponential of purely imaginary numbers*

$$r = \exp_{S^1}(i\varphi)$$

with $\varphi \in \mathbb{R}$. Expanding the exponential map into a Taylor series $\exp_{S^1}(x) = 1 + \frac{x}{1!} + \frac{x^2}{2!} + \frac{x^3}{3!} + \frac{x^4}{4!} + \ldots$ and inserting a complex number into the series leads to

$$\exp_{S^1}(i\varphi) = 1 + i\varphi - \frac{\varphi}{2!} + i\frac{\varphi}{3!} - \frac{\varphi}{4!} + \ldots .$$

Comparing the previous equation to the Taylor expansions of the trigonometric functions given by

$$\cos(\varphi) = 1 - \frac{\varphi^2}{2!} + \frac{\varphi}{1!} + \frac{\varphi^4}{4!} - \ldots \qquad \textit{and} \qquad \sin(\varphi) = \frac{\varphi}{1!} - \frac{\varphi^3}{3!} + \frac{\varphi^5}{5!} - \ldots .$$

reveals the following closed-form of the exponential map

$$r = \exp_{S^1}(i\varphi) = \cos(\varphi) + i\sin(\varphi)\,.$$

The logarithmic map is given by the logarithm of complex numbers. For a complex number $r = a + ib \in S^1$ *it is given by*

$$\log_{S^1}(r) = i\arctan\left(\frac{b}{a}\right) = i\varphi\,,$$

where $\log_{S^1}(r)$ *is uniquely defined for* $-\pi < \varphi \leq \pi$. *Both maps are visualized in Fig. 3.3.*

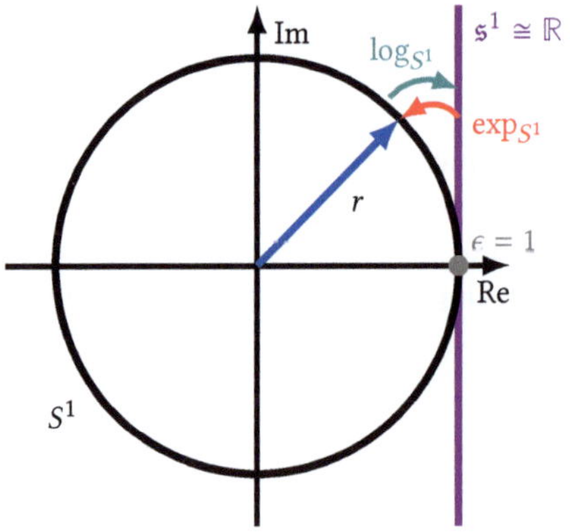

Figure 3.3.: Sketch of the unit sphere S^1 with Lie algebra.

A more abstract Lie group, which gained greater interest in recent years for the description of the geometrically exact beam, is the special Euclidean group SE(3) [66, 120, 134–136, 145].

Example 3.1.8 (Special Euclidean group $SE(3)$). *Motions of rigid bodies are usually expressed by position vector* $\boldsymbol{\varphi} \in \mathbb{R}^3$ *and a rotational tensor* $\mathbf{R} \in SO(3)$. *The solution is then sought in the direct product of both groups* $\mathbb{R}^3 \times SO(3)$, *which leaves the groups decoupled. An alternative way to express the motion is the special Euclidean group* $SE(3)$. *It is a semidirect product*[3] *given by* $SE(3) = \mathbb{R}^3 \rtimes SO(3)$, *which leads to a direct coupling of rotations and displacements. The group can be represented by tensors* $\mathbf{E}(\mathbf{R}, \boldsymbol{\varphi})$ *[66, 72, 136]*

$$SE(3) := \left\{ \mathbf{E} = \begin{bmatrix} \mathbf{R} & \boldsymbol{\varphi} \\ 0_{1\times 3} & 1 \end{bmatrix} \in \mathbb{R}^{4\times 4} \,\middle|\, \mathbf{R} \in \mathbb{R}^{3\times 3}, \quad \mathbf{R}^\top \cdot \mathbf{R} = \mathbf{I}, \quad \det(\mathbf{R}) = 1, \quad \boldsymbol{\varphi} \in \mathbb{R}^3 \right\}.$$

[3] A direct product group $\mathcal{A} = \mathcal{B} \times \mathcal{C}$ is a group, where the group action is defined through, $(b_1, c_1) \circ (b_2, c_2) = (b_1 \circ b_2, c_1 \circ c_2)$ where $b_1, b_2 \in \mathcal{B}$ and $c_1, c_2 \in \mathcal{C}$. A semidirect product group $\mathcal{D} = \mathcal{E} \rtimes \mathcal{F}$ is a group, where the group action is given by $(e_1, f_1) \circ (e_2, f_2) = (e_1 \circ f_1 \circ e_2, f_1 \circ f_2)$ with $e_1, e_2 \in \mathcal{E}$ and $f_1, f_2 \in \mathcal{F}$. [147, Chapter 10].

The corresponding group action is the matrix multiplication $(\cdot)$. *The identity element is the identity tensor* $\epsilon = \mathbf{1}_{4\times4}$ *and its inverse is given by* $\mathbf{E}^{-1}(\mathbf{R}, \boldsymbol{\varphi}) = \mathbf{E}(\mathbf{R}^\top, -\mathbf{R}\cdot\boldsymbol{\varphi})$. *The associated Lie algebra* $\mathfrak{se}(3)$ *is the space of* (4×4)*-tensors defined by [66, 72, 136]*

$$T_\epsilon SE(3) = \mathfrak{se}(3) := \left\{ \tilde{\mathbf{e}} = \begin{bmatrix} \hat{\mathbf{v}} & \mathbf{w} \\ \mathbf{0}_{1\times3} & 0 \end{bmatrix} \middle| \hat{\mathbf{v}} + \hat{\mathbf{v}}^\top = \mathbf{0}, \quad \mathbf{w} \in \mathbb{R}^3 \right\},$$

where $\hat{\mathbf{v}}$ *is a skew-symmetric tensor*[4]. *It, thus, can be deduced that the Lie algebra is six dimensional. The associate exponential map is given by [66, 136]*

$$\exp_{SE(3)}(\hat{\mathbf{e}}) = \begin{bmatrix} \exp_{SO(3)}(\hat{\mathbf{v}}) & \mathbf{T}^\top_{SO(3)}(\hat{\mathbf{v}})\cdot\mathbf{w} \\ \mathbf{0}_{1\times3} & 1 \end{bmatrix},$$

where $\exp_{SO(3)}$ *is the exponential map of the special orthogonal group, which is given below (see Section 3.2).* $\mathbf{T}_{SO(3)}(\hat{\mathbf{v}})$ *is the so-called tangent application of the special orthogonal group* $SO(3)$ *[66, 136], which is given by*

$$\mathbf{T}_{SO(3)}(\hat{\mathbf{v}}) = \mathbf{I} + \frac{1-\cos(\|\mathbf{v}\|)}{\|\mathbf{v}\|^2}\hat{\mathbf{v}} + \frac{\|\mathbf{v}\| - \sin(\hat{\mathbf{v}})}{\|\mathbf{v}\|^3}\hat{\mathbf{v}}^2,$$

where $\mathbf{v}$ *is the associated vector of* $\hat{\mathbf{v}}$. *The corresponding logarithmic map is given by [66]*

$$\log_{SE(3)}(\mathbf{E}) = \begin{bmatrix} \log_{SO(3)}(\mathbf{R}) & \mathbf{T}^{-\top}_{SO(3)}(\log_{SO(3)}(\mathbf{R}))\cdot\boldsymbol{\varphi} \\ \mathbf{0}_{1\times3} & 1 \end{bmatrix},$$

where $\log_{SO(3)}(\mathbf{R})$ *is the logarithmic map of the special orthogonal group* $SO(3)$, *which is given in Section 3.2.*

In the subsequent sections Section 3.2 and Section 3.3.3 the exponential and logarithm maps of the special orthogonal group and quaternions are introduced.

3.2. The special orthogonal group

It is well-known that the rotation in three-dimensional space is defined by a second-order tensor $\mathbf{R}$, which can be used to compute any rotation of an arbitrary vector $\mathbf{v} \in \mathbb{R}^3$ by

$$\mathbf{v}' = \mathbf{R}\cdot\mathbf{v}, \tag{3.8}$$

where $\mathbf{v}'$ is the rotated vector. A pure rotation does not change the length of the vector $\mathbf{v}$, which leads to conditions for $\mathbf{R}$. The following equation has to hold for $\mathbf{R}$ to conserve the length of $\mathbf{v}$

$$\mathbf{v}\cdot\mathbf{v} = \mathbf{v}'\cdot\mathbf{v}' = (\mathbf{R}\cdot\mathbf{v})\cdot(\mathbf{R}\cdot\mathbf{v}) = \mathbf{v}\cdot\mathbf{R}^\top\cdot\mathbf{R}\cdot\mathbf{v}. \tag{3.9}$$

[4] Compare to the Lie algebra of the special orthogonal group $SO(3)$ in Section 3.2.

The equation is fulfilled if

$$\mathbf{R}^\top \cdot \mathbf{R} = \mathbf{I}\,. \tag{3.10}$$

Thus, the inverse of the rotational tensor $\mathbf{R}^{-1}$ is given by its transpose $\mathbf{R}^\top$. Tensors fulfilling Eq. (3.10) are called orthogonal tensors. The group of orthogonal tensors is denoted by $O(3)$ and is a subgroup of the general linear group $GL(3)$ and thus a Lie group itself [72, Chapter 2]. To further ensure, that $\mathbf{R}$ defines a pure rotation and preserves the orientation we also need to ensure that

$$\det(\mathbf{R}) = 1 \tag{3.11}$$

holds. Both conditions, Eq. (3.10) and Eq. (3.11), give the equations defining the embedded nonlinear manifold

$$SO(3) := \{\mathbf{R} \in \mathbb{R}^{3\times 3} \,|\, \mathbf{R}^\top \cdot \mathbf{R} = \mathbf{I}, \quad \det(\mathbf{R}) = 1\}\,, \tag{3.12}$$

which defines the so-called special orthogonal tensors $\mathbf{R} \in SO(3)$. Often the term proper orthogonal tensor is used synonymously.

The special orthogonal group does not admit an abelian group structure. This can be illustrated by examining an object from daily life as shown in Fig 3.4. Two different rotation sequences of a dice with the same initial position are displayed. The first sequence, in Fig. 3.4a, is given as follows. First, a rotation with an angle of 90° is performed around the spatially fixed $\mathbf{e}_2$-axis. In the second step a rotation by −90° around the axis $\mathbf{e}_3$ is executed. Below, in Fig. 3.4b the rotation order is reversed. Clearly, both rotation sequences lead to different outcomes. Thus, the conclusion can be drawn that rotations in three dimensions are neither additive nor commutative.

Indeed the group's action is the scalar product between tensors $(\cdot)$, which is not commutative. As mentioned before, it is a subgroup of the general linear group, thus, the unit tensor $\mathbf{I} \in \mathbb{R}^{3\times 3}$ is the group's identity element $\epsilon = \mathbf{I}$, and the inverse given by the transpose $\mathbf{R}^{-1} = \mathbf{R}^\top$ as shown in Eq. (3.10). The corresponding Lie algebra $\mathfrak{so}(3)$ can be found by differentiating the group constraints in Eq. (3.10) with respect to time

$$\dot{\mathbf{R}}^\top \cdot \mathbf{R} + \mathbf{R}^\top \cdot \dot{\mathbf{R}} = \mathbf{0} \tag{3.13}$$

and evaluating the derivatives at the identity $\mathbf{R} = \epsilon$, which reveals

$$\dot{\mathbf{R}}^\top + \dot{\mathbf{R}} = \mathbf{0}\,. \tag{3.14}$$

It follows from Eq. (3.14) that tensors on the Lie algebra have to be skew-symmetric. Therefore, the Lie algebra is given by

$$T_\epsilon SO(3) = \mathfrak{so}(3) := \{\hat{\boldsymbol{\omega}} \in \mathbb{R}^{3\times 3} \,|\, \hat{\boldsymbol{\omega}} + \hat{\boldsymbol{\omega}}^\top = \mathbf{0}\}\,, \tag{3.15}$$

where a mapping $\hat{(\bullet)} : \mathbb{R}^3 \mapsto \mathfrak{so}(3)$ is introduced. The Lie algebra is isomorph to $\mathbb{R}^3$. The components of the corresponding vector of the skew-symmetric tensor $\hat{\boldsymbol{\omega}}$ can be computed from [95, Chapter 15]

$$\omega_i = \frac{1}{2}\epsilon_{ijk}\hat{\omega}_{kj} \tag{3.16}$$

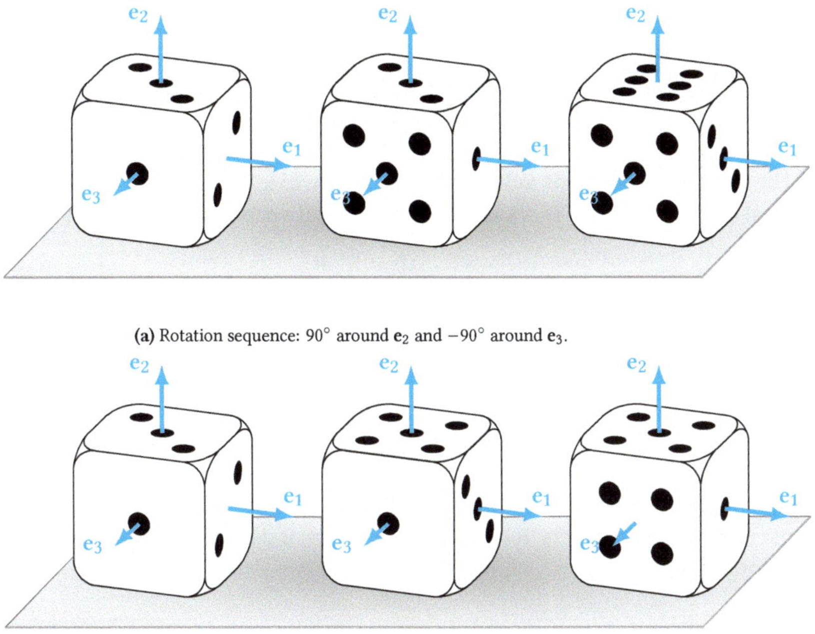

(a) Rotation sequence: 90° around $\mathbf{e}_2$ and −90° around $\mathbf{e}_3$.

(b) Rotation sequence: −90° around $\mathbf{e}_3$ and 90° around $\mathbf{e}_2$.

Figure 3.4.: Two different rotations sequences of a dice.

The vector $\boldsymbol{\omega}$ is referred to as the axial vector of $\hat{\boldsymbol{\omega}}$. Note that in this work we apply the Einstein notation for double indices, where Greek letters $\alpha, \beta \in 1, 2$ and lower Roman letter indices i, j, k range from 1 to 3.
As introduced above, Lie groups have a mapping, which maps elements from the Lie algebra to the manifold. For $SO(3)$ the mapping $\exp_{SO(3)} : \mathfrak{so}(3) \mapsto SO(3)$ is given by

$$\mathbf{R} = \exp_{SO(3)}(\hat{\boldsymbol{\omega}}) = \mathbf{I} + \frac{\sin(||\boldsymbol{\omega}||)}{||\boldsymbol{\omega}||}\hat{\boldsymbol{\omega}} + \frac{1 - \cos(||\boldsymbol{\omega}||)}{||\boldsymbol{\omega}||^2}\hat{\boldsymbol{\omega}}^2 , \tag{3.17}$$

where $\mathbf{R} \in SO(3)$ and $\hat{\boldsymbol{\omega}} \in \mathfrak{so}(3)$. The exponential map of the special orthogonal group is also often referred to as the Rodrigues' rotation formula. A derivation of Eq. (3.17) is given in [95, Chapter 9].
The logarithmic map $\log_{SO(3)} : SO(3) \mapsto \mathfrak{so}(3)$ is given by [66]

$$\log_{SO(3)}(\mathbf{R}) = \frac{\theta}{2\sin(\theta)}\left(\mathbf{R} - \mathbf{R}^\top\right) , \qquad \text{where} \qquad \theta = \arccos\left(\frac{1}{2}\operatorname{tr}(\mathbf{R}) - 1\right) . \tag{3.18}$$

An object from the Lie algebra $\hat{\boldsymbol{\Omega}} \in \mathfrak{so}(3)$ can be mapped onto the tangent space $T_{\mathbf{R}}SO(3)$ by

$$\dot{\mathbf{R}} = \mathbf{R} \cdot \hat{\boldsymbol{\Omega}} \qquad \in \qquad T_{\mathbf{R}}SO(3) , \tag{3.19}$$

which reveals the time derivative of a special orthogonal tensor. The tensor $\dot{\mathbf{R}}$ is skew-symmetric due to the mutual orthogonality of symmetric and skew-symmetric tensors [16, Chapter 1]. The corresponding vector $\boldsymbol{\Omega}$ from

$$\hat{\boldsymbol{\Omega}} = \mathbf{R}^\top \cdot \dot{\mathbf{R}} \tag{3.20}$$

is the convective angular velocity [95, Chapter 15]. Through a rotation of $\hat{\boldsymbol{\Omega}}$ by $\mathbf{R}$ the spatial angular velocity is obtained. It is the axial vector $\boldsymbol{\omega}$ of [95, Chapter 15]

$$\hat{\boldsymbol{\omega}} = \dot{\mathbf{R}} \cdot \mathbf{R}^\top \,. \tag{3.21}$$

3.3. Parametrizations of the special orthogonal group

Various options exist to represent tensors of $SO(3)$, which all have advantages and drawbacks when it comes to their application in a numerical method. In this section, we want to give a short overview of three possible approaches, without the claim to cover the topic in its entirety. A very theoretical discussion of different parametrizations is given by Stuelpnagel in [142].

3.3.1. Euler angles

Global rotational angles, such as Euler or Tait angles, are the parametrization engineers are usually most accustomed to. We focus here on the description of Euler angles. The basic idea of Euler angles is to compute the rotational tensor from a sequence of rotations. The total rotational tensor is obtained from

$$\mathbf{R} = \mathbf{R}_3 \cdot \mathbf{R}_2 \cdot \mathbf{R}_1 \,, \tag{3.22}$$

where each tensor $\mathbf{R}_i$, $i = 1, \dots, 3$, is a rotational tensor around a predefined axis. The rotation around only a single axis can be constructed with relative ease from geometrical considerations. There exist six possible combinations in which the rotation order can be defined. We use the definition given in [6, 95]. The total rotation given by $\mathbf{R}$ rotates the basis $\mathbf{e}_i$ onto a new basis $\mathbf{d}_i$. The first step of the rotation sequence is a rotation by an angle α_1 around axis $\mathbf{e}_3$ with the tensor $\mathbf{R}_1$

$$\mathbf{e}_i' = \mathbf{R}_1 \cdot \mathbf{e}_i = \begin{bmatrix} \cos(\alpha_1) & \sin(\alpha_1) & 0 \\ -\sin(\alpha_1) & \cos(\alpha_1) & 0 \\ 0 & 0 & 1 \end{bmatrix} \cdot \mathbf{e}_i \,, \tag{3.23}$$

which results in the new basis $\mathbf{e}_i'$. In the second step, we rotate the basis $\mathbf{e}_i'$ by an angle α_2 with $\mathbf{R}_2$ around the axis $\mathbf{e}_1'$

$$\mathbf{e}_i'' = \mathbf{R}_2 \cdot \mathbf{e}_i' = \begin{bmatrix} 1 & 0 & 0 \\ 0 & \cos(\alpha_2) & \sin(\alpha_2) \\ 0 & -\sin(\alpha_2) & \cos(\alpha_2) \end{bmatrix} \cdot \mathbf{e}_i' \,, \tag{3.24}$$

from which the basis $\mathbf{e}_i''$ is obtained. With the last rotation by the angle α_3 around the axis $\mathbf{e}_3''$ using tensor $\mathbf{R}_3$, we obtain the final basis $\mathbf{d}_i$

$$\mathbf{d}_i = \mathbf{R}_3 \cdot \mathbf{e}_i'' = \begin{bmatrix} \cos(\alpha_3) & \sin(\alpha_3) & 0 \\ -\sin(\alpha_3) & \cos(\alpha_3) & 0 \\ 0 & 0 & 1 \end{bmatrix} \cdot \mathbf{e}_i'' \,. \tag{3.25}$$

The set of unknowns is hereby given by the triple of rotational angles [95, Chapter 15]

$$Q_\alpha = \{(\alpha_1\,\alpha_2\,\alpha_3) \,|\, \alpha_1, \alpha_2 \in [0, 2\pi)\,;\alpha_3 \in [0, \pi)\}\,. \tag{3.26}$$

The computation of the rotational tensor is seemingly evident, however, the approach becomes difficult, by closer inspection. Even though, it is often spoken of as a rotational vector, the triple α_i does not lie in a vector space. Thus, many computations are complicated to perform. For example, due to their non-additive nature, the triple can in general not be discretized [11].
Furthermore, there exists no parametrization with three unknowns, which covers $SO(3)$ completely. A proof using topological arguments is given in [142]. Therefore, every parametrization of the special orthogonal group with three global coordinates exhibits singularities. In the case of Euler angles, this is known as Gimbal lock [95, Chapter 15]. It occurs if two of the axes of the bases $\mathbf{e}_i$ to $\mathbf{e}_j''$ for $i, j = 1, 2, 3$ coincide.

With only three degrees of freedom, Euler angles represent a minimal set of coordinates to describe the special orthogonal group. Thus, the use of global rotational vectors does not lead to a saddle point structure, which is favorable from a numerical viewpoint. But, as shown in the computation of the rotation tensor above many trigonometric functions have to be evaluated, which are numerically costly [30] and lead to lengthy expressions in the analytical analysis.

3.3.2. Directors

Besides parametrizations of the special orthogonal group, it is possible to use the components of a tensor $\mathbf{R} \in SO(3)$ directly as degrees of freedom in the design of a numerical method.
Betsch and Steinmann [21, 22] introduce this concept as the director formulation, where the orientation of a body is described by an orthonormal body-fixed frame $\mathbf{d}_i \in \mathbb{R}^3$, where the vectors

$$\mathbf{d}_i = \mathbf{R} \cdot \mathbf{e}_i \,. \tag{3.27}$$

are referred to as directors. Thus, the resulting set of coordinates describing the rotation is given by

$$Q_\mathrm{d} = \{\begin{bmatrix}\mathbf{d}_1 & \mathbf{d}_2 & \mathbf{d}_3\end{bmatrix} \in SO(3)\}\,. \tag{3.28}$$

The constraints of $SO(3)$ can be rewritten into a condition for the directors

$$\mathbf{d}_i \cdot \mathbf{d}_j = \delta_{ij}\,, \tag{3.29}$$

which ensures the normality and mutual orthogonality of the directors and consequently, in view of Eq. (3.27), $\mathbf{R} \in SO(3)$. Choosing the appropriate boundary conditions eliminates the necessity to enforce Eq. (3.11) in boundary values problems.

Similarly to Euler angels, the directors are also an intuitive and easily interpretable approach. Compared to the former, however, they have the advantage that no trigonometric functions, which are computationally costly, have to be evaluated in the numeric evaluation. Furthermore, the director formulation often simplifies the constructions of constraints [5]. Unfortunately, the number of coordinates is increased by factor three to a total of nine in comparison to the three minimally required coordinates. Furthermore, in a numerical framework, it becomes necessary to enforce the group constraints of $SO(3)$ (Eq. (3.29)). When enforcing the group constraint with Lagrange multipliers, this leads to a saddle point structure of the numerical formulation, which can lead to challenges in the numerical solution [15].

3.3.3. Unit quaternions

Another possibility to represent rotations is the use of quaternions, which are at first a rather abstract concept. The discovery of quaternions is mostly attributed to Sir William Hamilton, who first published his results in 1843, however, Olinde Rodrigues made the same discovery already three years earlier [64, Chapter 1]. Nevertheless, the space of quaternions is denoted by $\mathbb{H}$ in honor of Hamilton, who famously campaigned for their use. The campaign of Hamilton was not successful at first. During the 19th century, the concept of quaternions was met with a lot of skepticism as becomes, for example, obvious by a quote from Heaviside from 1983 [68, Chapter 3], which states: "*...the quaternion was not only not required, but was a positive evil of no inconsiderable magnitude...*". Several statements with a similar message can be found in the same source. Nevertheless, the need to efficiently represent rotations in many fields, from quantum mechanics up to computer graphics, led to the fact that quaternions are a well-established concept today [64, Chapter 1].
In the present work, we also consider quaternions and briefly introduce them in this section based on [20, 34, 36, 64] and use the notation proposed in [20].

3.3.3.1. Fundamentals of quaternions

Quaternions are often introduced as an extension of the complex numbers. Similar to complex numbers they consist of a scalar real part $q_0 \in \mathbb{R}$ and an imaginary or vector part $\mathbf{q}$

$$\mathbb{q} = q_0 + q_1\mathbf{i} + q_2\mathbf{j} + q_3\mathbf{k} = (q_0, \mathbf{q}) \tag{3.30}$$

[5] Many different formulations of joints for multibody kinematics in a director formulation are given in [124].

Quaternions can be multiplied by a scalar, added, subtracted, or multiplied with a scalar product in the same manner as vectors from $\mathbb{R}^4$. However, a new operation between two quaternions is defined by

$$\mathbb{q} \circ \mathbb{p} = (q_0 p_0 - \mathbf{q} \cdot \mathbf{p}, q_0 \mathbf{p} + p_0 \mathbf{q} + \mathbf{q} \times \mathbf{p}) \ . \tag{3.31}$$

The operator $(\cdot)$ denotes the scalar product and $(\times)$ is the cross product between vectors of $\mathbb{R}^3$. We call the operator $(\circ)$ the quaternion product or quaternion multiplication. In contrast to the scalar product between vectors the quaternion product of two quaternions $\mathbb{q}, \mathbb{p} \in \mathbb{H}$ is closed[6] as it results in a quaternion $\mathbb{q} \circ \mathbb{p} = \mathbb{v} \in \mathbb{H}$. Due to the last term in Eq. (3.31) the quaternion product is not commutative. Similar to imaginary numbers a conjugate of a quaternion is defined by the negative of the imaginary or vector part

$$\bar{\mathbb{q}} = (q_0, -\mathbf{q}) \ . \tag{3.32}$$

The 2-norm of a quaternion is defined by

$$||\mathbb{q}|| = \sqrt{\mathbb{q} \cdot \mathbb{q}} = \sqrt{q_0^2 + q_1^2 + q_2^2 + q_3^2} , \tag{3.33}$$

equivalent to vectors from $\mathbb{R}^4$. Note that the 2-norm is also referred to as length. Using Eq. (3.32) and Eq. (3.33) the inverse of a quaternion $\mathbb{q}^{-1}$ can be computed by

$$\mathbb{q}^{-1} = \frac{1}{||\mathbb{q}||^2} \bar{\mathbb{q}} , \tag{3.34}$$

where the inverse is defined by

$$\mathbb{q} \circ \mathbb{q}^{-1} = (1, \mathbf{0}) \ . \tag{3.35}$$

As mentioned above, quaternions are closed under the quaternion product. Thus, they form a group with quaternion product as group operation, the identity element $\epsilon = (1, \mathbf{0})$, and the inverse as defined in Eq. (3.35). Furhter, the quaternion product is associative $(\mathbb{q} \circ \mathbb{p}) \circ \mathbb{v} = \mathbb{q} \circ (\mathbb{p}) \circ) \mathbb{v}$.

3.3.3.2. Fundamentals of unit quaternions

Quaternions of unit length are of special interest for applications in engineering problems since they can be used to parametrize rotations in three dimensions. Unit quaternions form a parametrization of the unit sphere S^3 in $\mathbb{R}^4$

$$S^3 := \{\mathbb{q} \in \mathbb{H} \mid \mathbb{q} \cdot \mathbb{q} = 1\} , \tag{3.36}$$

[6] A proof that a triple is insufficient and indeed a quadruple is needed to define a rule, which is closed under multiplication is for example given in [34, Chapter 1].

which is a smooth manifold. Thus, they exhibit a Lie group structure, with the group properties introduced above. The unit constraint can be rewritten as

$$\mathbb{q} \cdot \bar{\mathbb{q}} = 1\,. \tag{3.37}$$

By taking the time derivative of the unit constraint in Eq. (3.37) we obtain

$$\dot{\mathbb{q}} \cdot \bar{\mathbb{q}} + \mathbb{q} \cdot \dot{\bar{\mathbb{q}}} = 0\,. \tag{3.38}$$

When evaluating Eq. (3.38) at the group's identity $\mathbb{q} = \epsilon$

$$\dot{\mathbb{q}} + \dot{\bar{\mathbb{q}}} = (0, \mathbf{0}) \tag{3.39}$$

we obtain the admissible velocities at $T_\epsilon S^3$ as pure quaternions. Pure quaternions have a scalar part equal to zero. The associated Lie algebra $\mathfrak{s}^3$ of the unit sphere S^3 is, thus, given by

$$T_\epsilon S^3 = \mathfrak{s}^3 := \{\mathbb{v} \in \mathbb{H} \mid \mathrm{Re}(\mathbb{v}) = v_0 = 0\}\,. \tag{3.40}$$

and is isomorph to $\mathbb{R}^3$. An object from the Lie algebra $\mathbb{v}$ can be mapped to the tangent space $T_\mathbb{q} S^3$ at $\mathbb{q}$ by applying the quaternion product

$$\dot{\mathbb{q}} = \mathbb{q} \circ \mathbb{v} \qquad \in T_\mathbb{q} S^3\,, \tag{3.41}$$

where $\mathbb{v} = (0, \frac{1}{2}\mathbf{\Omega})$. Hereby, the convective angular velocity $\mathbf{\Omega}$ is used. In terms of the spatial angular velocity $\boldsymbol{\omega}$ the admissible velocities can also be written as

$$\dot{\mathbb{q}} = \mathbb{w} \circ \mathbb{q} \qquad \in T_\mathbb{q} S^3\,, \tag{3.42}$$

where $\mathbb{w} = (0, \frac{1}{2}\boldsymbol{\omega})$. The factor of $1/2$ in both velocities is related to the quadratic terms of $\mathbb{q}$ in Eq. (3.46), which come into play, when taking the derivative. Elements from the Lie algebra can be mapped onto the manifold S^3 using the exponential map $\exp_{S^3} : \mathfrak{s}^3 \mapsto S^3$

$$\exp_{S^3}((0, \mathbf{q})) = \cos(\|\mathbf{q}\|)(1, \mathbf{0}) + \frac{\sin(\|\mathbf{q}\|)}{\|\mathbf{q}\|}(0, \mathbf{q})\,. \tag{3.43}$$

The inverse of the exponential map is given by the logarithm $\log_{S^3} : S^3 \mapsto \mathfrak{s}^3$, which can also be given in a closed form [91]

$$\log_{S^3}(\mathbb{q}) = 2\left(0, \arccos\left(\frac{q_0}{\|\mathbb{q}\|}\right)\frac{\mathbf{q}}{\|\mathbb{q}\|}\right)\,. \tag{3.44}$$

3.3.3.3. Parametrization of the special orthogonal group

Any rotation of a vector $\mathbf{v} \in \mathbb{R}^3$ can be expressed with the help of a unit quaternion $\mathbb{q} \in S^3$ via

$$\mathbb{q} \circ (0, \mathbf{v}) \circ \bar{\mathbb{q}} = (0, \mathbf{v}')\,, \tag{3.45}$$

where $\mathbf{v}' \in \mathbb{R}^3$ is the rotated vector. Even though the rotation of an element $\mathbf{v}$ of $\mathbb{R}^3$ can be written in terms of quaternion algebra it is more convenient to apply the mapping $\mathbf{R} : S^3 \mapsto SO(3)$ given by the Euler-Rodrigues parametrization [95, Chapter 9]

$$\mathbf{R}(\mathbb{q}) = (q_0^2 - \mathbf{q} \cdot \mathbf{q})\mathbf{I} + 2\mathbf{q} \otimes \mathbf{q} + 2q_0\hat{\mathbf{q}} \tag{3.46}$$

and apply the resulting rotational tensor in a standard fashion

$$\mathbf{R}(\mathbb{q}) \cdot \mathbf{v} = \mathbf{v}' . \tag{3.47}$$

Note that the mapping from the unit sphere S^3 onto $SO(3)$ is not unique. The same rotation tensor is described by $\mathbb{q}$ and $-\mathbb{q}$, which can be established from Eq. (3.46). Thus, there exists no explicit inverse mapping from $SO(3)$ to S^3. Numerical algorithms, such as Spurrier's algorithm [138], have to be used instead.

In contrast to other parametrization with rotational vectors, such as Euler angles, there are no computationally expensive trigonometric functions involved in Eq. (3.46). Furthermore, compared to the director approach the increase in the number of unknowns is moderate. Thus, a parametrization of the $SO(3)$ with quaternions is very efficient [30]. Additionally, there exist no singularities for a quaternion representation of the special orthogonal group. However, due to the unit length constraints, the use of quaternions results in a saddle point structure, when the constraint in Eq. (3.37) is enforced with a Lagrange multiplier.

3.4. Treatment of rotations in a finite element framework

When a problem is described using a suitable representation of rotations, the corresponding equations still need to be solved analytically or approximated using a numerical tool such as the FEM or IGA. At the heart of both methods, lies the discretization with a finite set of basis functions. Thus, we discuss here the treatment of rotations in the context of the FEM and IGA with appropriate discretization approaches.

Even though, the FEM and IGA are very well understood in most regards, dealing with complex nonlinear manifolds is a relatively new topic. Finding a suitable discretization approach for rotational fields on manifolds is one major issue. The discretization has to be frame-indifferent (or objective) and path-independent. Moreover, an optimal approach has to result in an optimal convergence behavior for h-refinement as well as p- and k-refinement.
In the case of the special orthogonal group, the problem was recognized early on [127, 128] yet a suitable approach was not found straight away. A first attempt was made by Simo and Qu-Voc [127, 128] to interpolate the incremental update of the rotation angle, while other authors used global rotation approaches e.g. [27, 81]. However, as shown by Crisfield and Jelenić [33] all these previously published approaches for the treatment of rotations in the context of the geometrically exact beam led to spurious results, because they were either

path-dependent, not frame-indifferent, or both. They proposed an alternative, which fulfills the requirements of frame-indifference and path-independence. More discussions on discretization approaches for rotations and motions are given in e.g. [11, 62, 114], where not all concepts lead to appropriate approaches.

Despite all the research effort put into the topic, it is still covered rather poorly in many regards. In particular, there are not many publications, which concern higher-order shape functions for the discretization of manifolds [122]. Publications, covering discretization approaches with higher continuity over elements in the context of the IGA, are especially rare.
Two approaches found in the literature for the general discretization of nonlinear manifolds are geodesic finite elements [120–122] and projection-based finite elements [49, 59, 102]. These two options and a classical, additive discretization approach are discussed in more detail. As a fourth option, we propose Gauss-Lobatto-based shape functions, which can only be applied to the classical FEM and not to the IGA.
The discretization approaches are discussed in detail for the discretization of directors and unit quaternions. As mentioned before, the discretization of global rotational angles, such as Euler angles, is not feasible [11] and, thus, not discussed here. Since a beam is a one-dimensional object, we limit the discussion here to a one-dimensional, single element with the coordinate $s \in [0, 1] = \Omega$.

3.4.1. Classical finite elements

As introduced in every textbook on the topic of finite elements, e.g. [77], the classical approach for an arbitrary discretized field $\mathbf{u}^h$ is a multiplication of the nodal or control point values with their corresponding shape function followed by a summation

$$\mathbf{u}^h(s) = \sum_{i=1}^{n} N_i(s)\mathbf{u}_i \,. \tag{3.48}$$

The shape functions are denoted by $N_i(s)$ for a FEM as well as an IGA approach. The weights $\mathbf{u}_i$ are either the values of the nodes or control points. In the sequel, we shall refer to the i-th point of the discretization as the i-th control/nodal point.
In this work, we call these elements classical finite elements since it is the standard approach used by the FEM as well as the IGA to approximate continuous fields, which lie in a flat space. When this approach is used without care for variables lying on a curved manifold, even higher-order elements cannot represent the curved manifold exactly.
In the case of a discretization approach, which fulfills the collocation properties at the nodal points, such as Lagrangian elements, the underlying geometry of the manifold can be preserved on these collocation points e.g. with the use of Lagrange multipliers. However, elsewhere in the domain, the geometry of the manifold is violated. This is especially relevant for the values at the integration points since the discretized field $\mathbf{u}^h$ is evaluated there in the numerical integration process.

3.4.1.1. Director approach

Neglecting the curvature of the underlying space of the special orthogonal group the discretization of the directors can be written as

$$\mathbf{d}_i^h(s) = \sum_{j=1}^{n} N_j(s)\mathbf{d}_i^j\,, \tag{3.49}$$

where $\mathbf{d}_i^j$ are the control/nodal point values of the directors. However, even if the orthonormality constraints are enforced in a strong sense on the nodes for the FEM, this does not lead to an orthonormal frame on the integration points. Thus, the underlying geometry of $SO(3)$ is not conserved by the discretization, i.e.

$$\mathbf{d}_i^h(s) \cdot \mathbf{d}_j^h(s) \neq \delta_{ij}\,. \tag{3.50}$$

Consequently, the discretization no longer exactly represents a rotational tensor on the level of the integration points. For higher orders this leads to a reduction of the convergence order as demonstrated in Chapter 5.

3.4.1.2. Unit Quaternions

A classical discretization approach for quaternions results in

$$\mathbb{q}^h(s) = \sum_{i=1}^{n} N_i(s)\mathbb{q}_i\,, \tag{3.51}$$

where $\mathbb{q}_i$ are the control/nodal point values of the quaternions. It is obvious that a classical discretization approach does not conserve the length of unit quaternions

$$\left\|\mathbb{q}^h(s)\right\| \neq 1\,. \tag{3.52}$$

Even if the control/nodal point values lie on the unit sphere S^3 itself the discretized approach does not represent the geometry of S^3 in the entire domain.
A sketch of the discretization approach in Fig. 3.5 elucidates the problem. The unit sphere S^3 is a four-dimensional object so it cannot be drawn. Thus, the sketch is reduced to two dimensions, and for simplicity, a linear interpolation is used. The control/nodal point values in blue lie on S^3. The discretization in between given by $\mathbb{q}^h(s)$ is clearly not on the unit sphere. Even higher-order approximation approaches are not able to approximate the curvature exactly. Even though the sketch exaggerates the error, and the use of a finer mesh leads to a better approximation, examples from in the subsequent Chapter 6 show, that the optimal convergence order is reduced by this approach for higher-order elements.

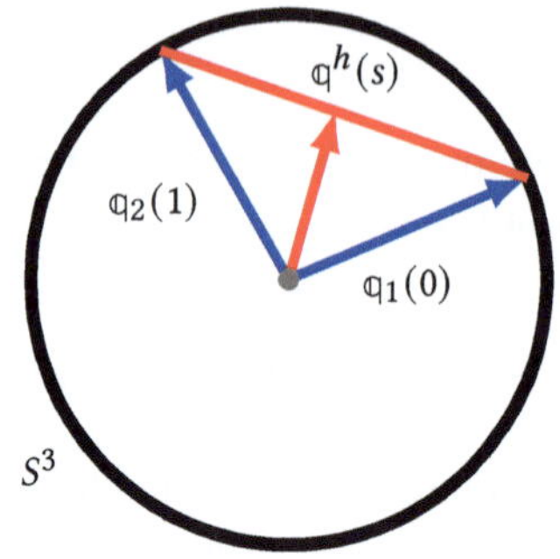

Figure 3.5.: Sketch of a classical finite element discretization of unit quaternions on the sphere S^3.

3.4.2. Geodesic finite elements

A generalized concept to interpolate manifolds is the use of geodesic finite elements. In the field of computational mechanics Crisfield and Jelenić [33, 83] were the first to propose an approach for rotational tensors, which falls into this broader concept of geodesic elements. However, in computer animations equivalent approaches were already proposed much earlier. In 1985 such an approach for unit quaternions was published by Shoemake [125]. In computational graphics, this approach is well-known and referred to as spherical linear interpolation (SLERP). A very similar approach can also be found in [87, 88] for B-Splines approximations of higher continuity [7].

However, the generalized interpolation approach under the name geodesic finite elements for arbitrary Riemannian manifolds was only introduced at the beginning of the last decade by Sander [120]. In [120] these elements were proposed as an equivalent of linear elements and only for one-dimensional domains. Geodesic elements were further advanced to higher dimensional elements [121] and for higher order Lagrange shape functions [122]. The corresponding mathematical proofs showing the optimality of the error bounds for the proposed element formulation were given in [58, 120]. Furthermore, an equivalent approach was used in combination with B-Spline shape functions in [60]. Geodesic elements were used in numerical examples, where in [120] they were applied to a formulation of the geometrically exact beam, and in [123] discussed in the context of Cosserat shells. Müller and Bischoff [102] apply the approach in combination with higher-order NURBS functions to the simulation of the Reissner-Mindlin shell model.

Geodesic elements on a manifold M are in general obtained by solving the following minimization problem

$$\mathbf{u}^h(s) = \arg\min_{\mathbf{u}\in M} \sum_{i=1}^{n} N_i(s)\,\mathrm{dist}(\mathbf{u}_i, \mathbf{u})^2\,, \tag{3.53}$$

[7] The SLERP interpolation is constructed along the geodesic, thus, they result in a constant norm of the velocity, which is of great advantage for animation [64].

where dist(•, •) is the intrinsic distance measure of the manifold M [59]. They, thus, follow from the minimization of the weighted distance, where the weights at s are given through the shape functions N_i. Using the intrinsic distance measure has many advantages. The geodesic elements are by design objective [59, 122] and due to the intrinsic construction derivatives and variations lie in the tangent space of the discretized manifold [59]. For arbitrary manifolds, the element formulation cannot be given in a closed form but has to be evaluated numerically [120]. Therefore, the geodesic finite elements are computationally expensive (cf. [102]), which especially limits their usability in dynamic simulations.

Using the knowledge of Lie groups from above, allows us to construct geodesic elements in a closed-form for matrix Lie groups. The discretization approach is given as

$$\mathbf{X}^h(s) = \mathbf{X}_1 \circ \exp_{\mathcal{G}}\left(\sum_{i=1}^{n} N_i(s)\tilde{\mathbf{v}}_i\right), \tag{3.54}$$

where $\mathbf{X}^h, \mathbf{X}_1, \mathbf{X}_i \in \mathcal{G}$ and $\tilde{\mathbf{v}}_i \in \mathfrak{g}$. Eq. (3.54) is a generalization of the approach given in [33], where the approach is limited to proper orthogonal tensors. Applications to Lie groups besides $SO(3)$ can already be found in the literature. In [143] the formulation is used for unit quaternions, and in [136] for the special Euclidean group. These elements are geodesic elements in the sense that the relative difference between the control/nodal points is computed with the intrinsic difference measure on the Lie group, which also minimizes the geodesic distance.
The control/nodal point values are $\mathbf{X}_1$ and $\mathbf{X}_i$. The element $\tilde{\mathbf{v}}_i$ describes the relative intrinsic difference between both control/nodal point values on the Lie algebra. It can be computed using the logarithmic map via

$$\tilde{\mathbf{v}}_i = \log_{\mathcal{G}}(\mathbf{X}_1^{-1} \circ \mathbf{X}_i), \tag{3.55}$$

where the index i ranges from 1 to n, which is the number of control/nodal points. Therein, $\tilde{\mathbf{v}}_i$ is computed between the i-th control/nodal point value $\mathbf{X}_i$ and the first control/nodal point value $\mathbf{X}_1$. Thus, for $i = 1$ there is no difference and, thus, $\tilde{\mathbf{v}}_1 = \mathbf{0}$. Different positions on the element can be used as reference instead of the value of the first control/nodal point $\mathbf{X}_1$, where various options are discussed in [33]. For simplicity, we choose here $\mathbf{X}_1$. As shown in [33], the use of the relative difference is necessary to achieve frame-indifference.

This approach discretizes the relative distance of the control/nodal point values on the Lie algebra, which is isomorph to a vector space. Thus, here a classical discretization can be applied.
Limitations on the element length are given through the interval, on which the logarithmic map is uniquely defined. If the relative distance between the control/nodal points is outside the interval a finer mesh needs to be chosen [33].
For the IGA this leads, in general, to a requirement of p-refinement with C^0-continuity over element boundaries, where Eq. (3.54) is applied on each individual element. This is due to the fact, that on the reference point $\mathbf{X}_1$ the shape functions need to fulfill the

collocation, since otherwise a coupling of the degrees of freedom over the entirety of the patch can occur. This is the case for the closed-form expression presented in [87, 88]. Implicit algorithms, as applied in [102], which minimizes the weighted distance in Eq.(3.53), can be used to achieve a discretization approach with higher continuity over a whole patch. Due to the local support property of the NURBS functions the distance in Eq. (3.53) only needs to be evaluated locally on the patch and not for all shape functions and control points simultaneously. Thus, a classical element-wise structure is obtained. To apply such an approach, a distance measure, intrinsic to the individual Lie group, is necessary. Derivatives of the shape functions can be obtained from the minimization problem [102, 122].

3.4.2.1. Director approach

Since the director approach is equivalent to a direct use of the rotational tensor, a geodesic approach for their discretization is given in [33]. On one single element, the discretization is given as

$$\mathbf{R}^h(s) = \mathbf{R}_1 \cdot \exp_{SO(3)}\left(\sum_{i=1}^{n} N_i(s)\hat{\boldsymbol{\omega}}_i\right), \tag{3.56}$$

where $\mathbf{R}_1$ is the control/nodal point value at $s = 0$. The skew-symmetric tensor $\hat{\boldsymbol{\omega}}_i$ is an element of the Lie algebra. It defines the relative rotation from $\mathbf{R}_i$ to $\mathbf{R}_1$. The rotation in between the control/nodal point values can be computed by applying the logarithmic map

$$\log_{SO(3)}(\mathbf{R}_1^\top \cdot \mathbf{R}_i) = \hat{\boldsymbol{\omega}}_i\,. \tag{3.57}$$

Due to the limits of the logarithmic map, the largest relative rotational angle cannot be bigger than π since otherwise $\hat{\boldsymbol{\omega}}_i$ is not uniquely defined [33].
The geodesic approach ensures, that the discretization results in a rotational tensor at every point in the finite element, i.e.

$$\mathbf{R}^h(s) \in SO(3) \quad \forall s \in \Omega\,. \tag{3.58}$$

The discretized directors can be computed from

$$\mathbf{d}_i^h(s) = \mathbf{R}^h(s) \cdot \mathbf{e}_i\,. \tag{3.59}$$

As a result, the geodesic approach ensures that the orthonormality constraint in Eq. (3.29) is fulfilled on every point of the domain

$$\mathbf{d}_i^h(s) \cdot \mathbf{d}_j^h(s) = \delta_{ij} \quad \forall s \in \Omega\,. \tag{3.60}$$

For the application of Eq. (3.53) in an implicit approach the minimal or geodesic distance measure of $SO(3)$ is needed. It is given by

$$\operatorname{dist}(\mathbf{R}_a, \mathbf{R}_b) = \left\|\log(\mathbf{R}_a^\top \cdot \mathbf{R}_b)\right\|_F, \tag{3.61}$$

where $\mathbf{R}_a, \mathbf{R}_b \in SO(3)$ [137]. The subscript F, hereby, indicates the use of the Frobenius norm given by

$$||\mathbf{A}||_F = \sqrt{\operatorname{tr}(\mathbf{A}^\top \cdot \mathbf{A})} \tag{3.62}$$

for a tensor $\mathbf{A} \in \mathbb{R}^{3\times 3}$.

3.4.2.2. Unit Quaternions

As mentioned above, in the computer graphics community the SLERP interpolation for unit quaternions was already proposed by Shoemaker [125] in 1985. To the best knowledge of the author, Ghosh and Roy [51] were the first to use this approach in combination with unit quaternions in computational mechanics for the beam, where they use an explicit formulation. In the context of the IGA it is employed by Tasora *et al.* [143], where, however, no detailed discussion concerning the discretization is given.

Equivalent to elements on $SO(3)$ the quaternion elements can be constructed by using the exponential map

$$\mathbb{q}^h(s) = \mathbb{q}_1 \circ \exp_{S^3}\left((0, \sum_{i=1}^{n} N(s)\mathbf{v}_i)\right), \tag{3.63}$$

where $\mathbb{q}_1$ is the control/nodal point value at $s = 0$. As before, $\mathbf{v}_i$ is given by the rotation between control point/nodal values $\mathbb{q}_i$ and $\mathbb{q}_1$. Thus, it can be computed by applying the logarithmic map

$$\log_{S^3}(\bar{\mathbb{q}}_1 \circ \mathbb{q}_i) = (0, \mathbf{v}_i)\,. \tag{3.64}$$

Through the construction of the elements using the geometry of the manifold, the discretization $\mathbb{q}^h(s)$ lies directly on the manifold S^3 at all points. The unit constraints are, therefore, fulfilled on the entire domain, i.e.

$$\left\|\mathbb{q}^h(s)\right\| = 1 \quad \forall s \in \Omega\,. \tag{3.65}$$

In a very similar manner to the geodesic distance measure of $SO(3)$, the geodesic distance for unit quaternions follows from

$$\operatorname{dist}(\mathbb{q}_a, \mathbb{q}_b) = \left\|\log_{S^3}(\bar{\mathbb{q}}_a \circ \mathbb{q}_b)\right\| \tag{3.66}$$

with $\mathbb{q}_a, \mathbb{q}_b \in S^3$.

An illustration of the approach is depicted in Fig. 3.6 for one single geodesic element. Similar to Section 3.4.1.2 we display only a first-order element with two control/nodal points. Due to the approach with the geodesic elements, the geodesic, the shortest distance on the curved manifold, connects both nodes. This is equivalent to a linear element in a flat space [120]. As a consequence of the intrinsic nature of the approach, the discretization always lies on the manifold.

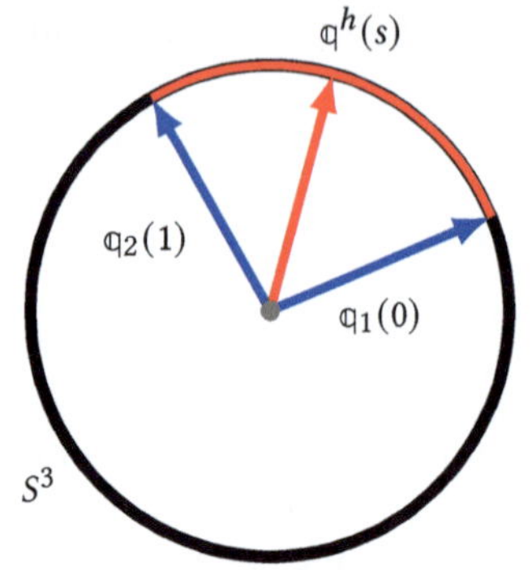

Figure 3.6.: Sketch of a geodesic finite element discretization of the unit sphere S^3

3.4.3. Projection-based finite elements

The third approach for the discretization of nonlinear manifolds discussed in this work is the employment of so-called projections-based finite elements. Various finite element formulations, which can be summarized under this concept, can be found in the literature. Romero [114] uses a discretization approach with the normalization of unit quaternions for the beam. Bauchau and Han [11, 62] discuss discretization approaches for different parametrizations of rotations and motions, which fall into this category. However, mathematical proofs are not given. In [102] projection-based finite elements are applied to the discretization of shells with NURBS basis functions.
In a more general approach, the concept was very recently proposed by Gawlik and Leok [49] and used for the special orthogonal group and unit quaternions. Grohs *et al.* [59] provides a further generalization to arbitrary manifolds as well as an analytical proof for the optimal error bounds.

Again, the goal of projection-based elements is to discretize quantities, which lie on a manifold $\mathcal{M}$. As given in [59], the element construction can be divided into two steps. First, an additive, classical discretization is performed

$$\mathbf{v}^h(s) = \sum_{i=1}^{n} N_i(s)\mathbf{v}_i\,, \tag{3.67}$$

where $\mathbf{v}_i \in \mathcal{M}$ are the control/nodal point values and N_i the employed shape functions, respectively. As discussed before, the discretized values do not lie on the manifold $\mathbf{v}^h(s) \notin \mathcal{M}$. Therefore, an orthogonal projection onto the manifold $\mathcal{M}$

$$\mathbf{u}^h(s) = \arg\min_{\mathbf{u}\in\mathcal{M}} \left\| \mathbf{u} - \sum_{i=1}^{n} N_i \mathbf{v}_i \right\|_{\mathbb{R}^n}^2 = \mathbf{P}_{\mathcal{M}}(\mathbf{v}^h) \tag{3.68}$$

is performed in a second step, where $||\bullet||_{\mathbb{R}^n}$ denotes the norm of the embedding space $\mathbb{R}^n$ and $\mathbf{P}_{\mathcal{M}}$ is the closest-point or orthogonal projection onto the manifold $\mathcal{M}$. The projection is well-defined close to the manifold [59]. The formulation is objective under rigid body motions if the shape functions form a partition of unity [59]. Thus, the projection-based approach can be applied to either the FEM or IGA without any further adaptations.

Since the projection-based elements are not constructed on the manifold itself, greater care has to be taken, when evaluating the derivatives and variations, respectively. Taking the derivative with respect to time of Eq. (3.68) results in [59]

$$\dot{\mathbf{u}}^h(s) = \mathbf{P}'_{\mathcal{M}}(\mathbf{v}^h) \cdot \sum_{i=1}^{n} N_i \dot{\mathbf{v}}_i \tag{3.69}$$

The derivative of the projector results in an orthogonal projector onto the tangent space $T_{\mathbf{u}}\mathcal{M}$ up to a Lipschitz constant [59], so that

$$\mathbf{P}'_{M}(\mathbf{v}^h) \rightarrow \mathbf{P}_{T_{\mathbf{u}}\mathcal{M}}(\mathbf{v}^h) \quad \text{for} \quad \left\| \mathbf{u} - \sum_{i=1}^{n} N_i \mathbf{v}_i \right\|^2_{\mathbb{R}^n} \rightarrow 0\,. \tag{3.70}$$

Therein, $\mathbf{P}_{T_{\mathbf{u}}\mathcal{M}}$ is the orthogonal projector onto the tangent space $T_{\mathbf{u}}\mathcal{M}$. Such a projection onto the tangent space can be used to obtain derivatives on a manifold. The application of such projectors is covered in depth in [2] and many projectors for various manifolds are given in [1]. Additional information on projectors is given in the Appendix A.

3.4.3.1. Director approach

Again we discuss the director approach in detail. We begin with a classical discretization approach

$$\mathbf{t}_i^h(s) = \sum_{j=1}^{n} N_j(s)\, \mathbf{t}_i^j\,, \tag{3.71}$$

where $\mathbf{t}_i^j \in \mathbb{R}^3$ are the control/nodal point values of the director frame. By assembling the discretized directors we obtain a tensor

$$\mathbf{T}^h(s) = \mathbf{t}_i^h(s) \otimes \mathbf{e}_i \tag{3.72}$$

Therein, $\mathbf{T}^h(s)$ is not a proper orthogonal tensor for all $s \in \Omega$. Thus, following the idea of the projection-based elements, an orthogonal projection onto the special orthogonal group is necessary. As shown in [108], the orthogonal projection onto $SO(3)$ is given by the polar decomposition

$$\mathbf{T}^h = \mathbf{R}^h \cdot \mathbf{U}^h\,, \tag{3.73}$$

where $\mathbf{R}^h$ is a proper orthogonal tensor and $\mathbf{U}^h$ accounts for the stretch of the discretization from the exact geometry of $SO(3)$. The polar decomposition can be computed numerically, which was done in [59]. Since it leads to very lengthy expressions, it is not practical to compute in a closed analytical form. From Eq. (3.73) the special orthogonal tensor follows from

$$\mathbf{R}^h = \mathbf{T}^h \cdot \left(\mathbf{U}^h\right)^{-1} = \mathbf{P}_{SO(3)}(\mathbf{T}^h)\,, \tag{3.74}$$

where $\mathbf{P}_{SO(3)}$ is the orthogonal projector onto $SO(3)$. After $\mathbf{R}^h$ is computed the individual directors are obtained from

$$\mathbf{d}_i^h = \mathbf{R}^h \cdot \mathbf{e}_i\,. \tag{3.75}$$

Since $\mathbf{R}^h$ is a proper orthogonal tensor, the orthonormality conditions are fulfilled in the whole domain

$$\mathbf{d}_i^h(s) \cdot \mathbf{d}_j^h(s) = \delta_{ij} \quad \forall \quad s \in \Omega\,. \tag{3.76}$$

When taking the derivative of Eq. (3.74) with respect to time the polar decomposition has to be taken into account, which leads to [29]

$$\begin{aligned}\dot{\mathbf{R}}^h(s) &= \mathbf{P}'_{SO(3)} : \dot{\mathbf{T}}^h \\ &= \frac{1}{\det(\mathbf{Z})}\mathbf{R}^h \cdot \mathbf{Z} \cdot \left(\left(\mathbf{R}^h\right)^\top \cdot \dot{\mathbf{T}}^h - \left(\dot{\mathbf{T}}^h\right)^\top \cdot \mathbf{R}^h\right) \cdot \mathbf{Z}\,,\end{aligned} \tag{3.77}$$

where

$$\mathbf{Z} = \operatorname{tr}\left(\mathbf{U}^h\right)\mathbf{I} - \mathbf{U}^h\,, \tag{3.78}$$

and the derivative of $\mathbf{T}^h$ with respect to time is given by

$$\dot{\mathbf{T}}^h = \sum_{j=1}^{n} N_j \dot{\mathbf{t}}_i^j \otimes \mathbf{e}_i\,. \tag{3.79}$$

In Eq. (3.77) the projector onto the tangent space $T_R SO(3)$ is contained, which is given by [2]

$$\mathbf{P}_{TSO(3)}(\mathbf{R}^h, \dot{\mathbf{T}}^h) = \frac{1}{2}\mathbf{R}^h \cdot \left(\left(\mathbf{R}^h\right)^\top \cdot \dot{\mathbf{T}}^h - \left(\dot{\mathbf{T}}^h\right)^\top \cdot \mathbf{R}^h\right)\,. \tag{3.80}$$

For $\mathbf{U}^h \to \mathbf{I}$ the derivative in Eq. (3.77) converges towards the projection given in Eq. (3.80). In [11] other possibilities than the polar decomposition as projection method are discussed. However, as the polar decomposition is the orthogonal projection the other approaches do not fall into the category of the projection-based approach.
Since the polar decomposition needs to be evaluated at every integration point in each iteration, the numerical cost for this approach is high because it involves a computationally expensive eigenvalue problem. Grohs *et al.* [59] report approximately the same computational time as for geodesic finite elements.

3.4.3.2. Unit Quaternions

In a finite element context, a projection-based discretization approach for unit quaternions in S^3 is given in [11, 114]. At first, the control/nodal values $\mathbb{p}_i$ of the quaternions are discretized in an additive manner

$$\mathbb{p}^h(s) = \sum_{i=1}^{n} N_i(s)\mathbb{p}_i \,. \tag{3.81}$$

Since $\mathbb{p}^h(s)$ does not have a unit length for every s, in a second step a closest-point projection to the manifold S^3 is applied. The closest-point projection is given by a normalization, resulting in the following discretized unit quaternion

$$\mathbb{q}^h(s) = \frac{\mathbb{p}^h(s)}{\left\|\mathbb{p}^h(s)\right\|} = \mathbf{P}_{S^3}(\mathbb{p}^h(s)) \,, \tag{3.82}$$

where $\mathbf{P}_{S^3}$ is the closest-point or orthogonal projector onto S^3. This approach ensures that the unit length is fulfilled at every point

$$\left\|\mathbb{q}^h(s)\right\| = 1 \quad \forall \quad s \in \Omega \,. \tag{3.83}$$

As above, the projection operator $\mathbf{P}_{S^3}$ has to be taken into account when taking the partial derivative of Eq. (3.82) with respect to time. It is given by

$$\dot{\mathbb{q}}^h = \mathbf{P}'_{S^3}(\mathbb{p}^h) \cdot \dot{\mathbb{p}}^h \,, \tag{3.84}$$

where the gradient of the projector is given by

$$\mathbf{P}'_{S^3}(\mathbb{p}^h) = \frac{1}{\left\|\mathbb{p}^h\right\|}\left(\mathbf{I}_4 - \mathbb{q}^h \otimes \mathbb{q}^h\right) \,. \tag{3.85}$$

The gradient of the projector can be recast in

$$\mathbf{P}'_{S^3}(\mathbb{p}^h) = \frac{1}{\left\|\mathbb{p}^h\right\|}\mathbf{P}_{TS^3}(\mathbb{q}^h) \,, \tag{3.86}$$

where we introduced a new projector $\mathbf{P}_{TS^3}$, which is a projector onto the tangent space $T_{\mathbb{q}}S^3$ [2]. The gradient $\mathbf{P}'_{S^3}$ convergence towards the projector $\mathbf{P}_{TS^3}$, if $\left\|\mathbb{p}^h\right\| \to 1$. Thus, the derivatives lie in the correct space and we fulfill the orthogonality requirement given in [11]

$$\mathbb{q}^h \cdot \dot{\mathbb{q}}^h = 0 \,, \tag{3.87}$$

if the projection-based element approach is applied.

A sketch of the projection-based approach for unit quaternions is drawn in Fig. 3.7. At first, the classical discretization is applied to obtain $\mathbb{p}^h(s)$, shown in gray. The subsequent normalization in Eq. (3.82) leads to a projection onto the unit sphere, such that $\mathbb{q}^h(s)$ is of unit length for all points in Ω.
The projection-based approach for unit quaternions is used in Chapter 6, where results are shown and discussed in detail.

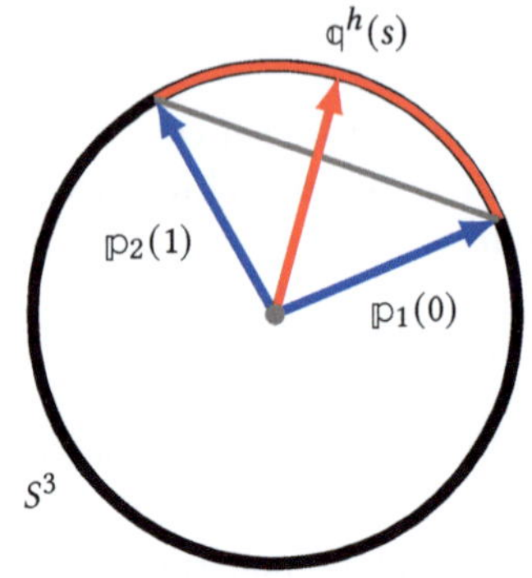

Figure 3.7.: Sketch of a projection-based finite element discretization of the unit sphere S^3.

3.4.4. Elements with Gauss-Lobatto-based shape functions

An approach that can only be applied for interpolatory approaches like the FEM and not the IGA is the use of Gauss-Lobatto-based shape functions. Even though, this approach relies on the additive approach, which was introduced as 'classical finite elements', we discuss it separately to avoid confusion. Similar to many topics in science, the naming of this approach is ambiguous. We use the name Gauss-Lobatto-based shape functions in accordance with [100]. To the best knowledge of the author, the utilization of Gauss-Lobatto-based shape functions has so far not been discussed in the context of interpolation on manifolds.

Like most standard approaches in the FEM, the Gauss-Lobatto-based shape functions use Lagrange basis functions, which are computed from [77]

$$N_i(\xi) = \prod_{j=1, i \neq j}^{p+1} \frac{\xi - \xi_j}{\xi_i - \xi_j} \tag{3.88}$$

for an interpolation of order p. However, they differ in the placement of the nodes ξ_i on the reference element. The outer nodes are placed at

$$\xi_1 = -1 \qquad \text{and} \qquad \xi_{p+1} = 1\,, \tag{3.89}$$

equivalent to the standard Lagrange shape functions. While the standard approach chooses equidistant nodes in the interior, for the Gauss-Lobatto-based shape functions the $p-1$ inner nodes are placed at the roots of the first derivative of the Legendre polynomial P_p of order p [133, Chapter 1]

$$\frac{\partial}{\partial \xi} P_p(\xi_i) = 0 \qquad \text{for} \quad i = 2, \ldots, p\,. \tag{3.90}$$

Even though the Gauss integration is the standard quadrature scheme in FEM other quadrature schemes can be employed, such as the Lobatto quadrature. The Lobatto quadrature uses the Gauss-Lobatto points as integration points. Compared to the Gauss integration, however, the Lobatto integration is not optimal, since it only integrates polynomials of order $p = 2n_{\mathrm{GL}} - 3$ exactly, where n_{GL} are the number of integration points. A detailed description of the Lobatto integration scheme is given in [133, Chapter 4].

If now the Gauss-Lobatto-based shape functions are combined with the Lobatto integration, the integration points are identical to the nodal positions. Therefore, using an enforcement of the group constraints in a pointwise or strong sense with Lagrange multipliers on the nodes leads to a satisfaction of the constraints at the integration points. Thus, when discretizing a nonlinear manifold using strong enforcement on the nodes in combination with the Gauss-Lobatto-based shape functions and the Lobatto quadrature, the discretized field is evaluated on the manifold's geometry. However, this leads to a strict coupling of the number of integration points and element order. Therefore, reduced integration approaches cannot be applied.

Fig. 3.8 highlights the differences between Lagrange and Gauss-Lobatto-based shape functions, where both approaches are displayed over a reference element. The displayed shape functions are cubic. Small differences between the shape functions themselves are visible, however, for elements of higher order $p > 4$, the differences become more apparent (cf. [55]). However, the nodal positions in Fig. 3.8a and Fig. 3.8b differ, where

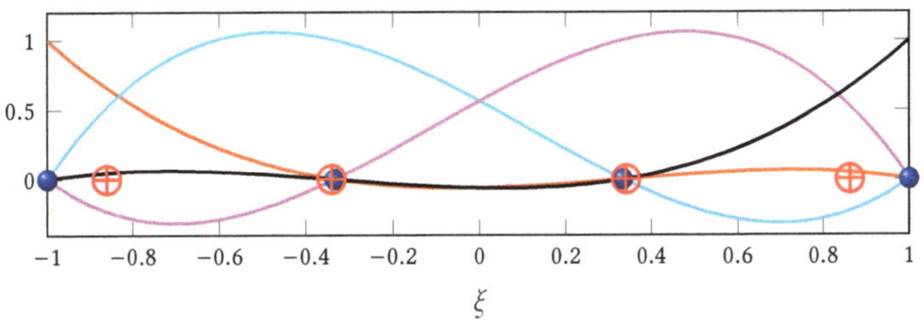

(a) Cubic Lagrange shape functions with equally space nodes and Gauss integration points.

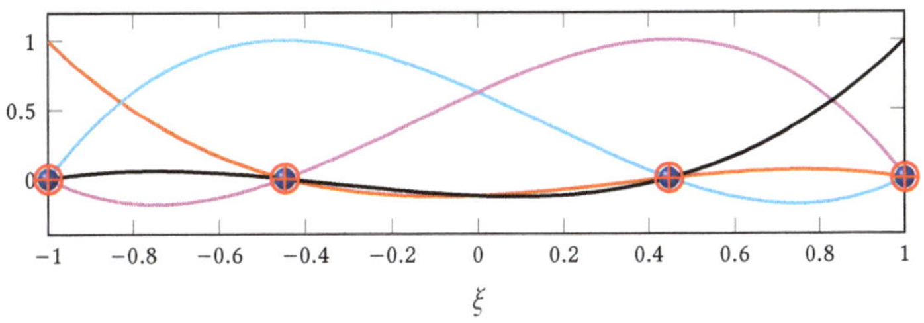

(b) Cubic Gauss-Lobatto-based shape functions with nodes on the Gauss-Lobatto points and Lobatto integration.

Figure 3.8.: Comparison between Lagrange and Gauss-Lobatto-based shape functions. The blue dots indicate the position of the nodes and the red cross-hairs display the positions of the integration points.

in Fig. 3.8 the blue dots indicate the position of the finite element nodes. The interior nodes using a standard approach with equidistant nodes are $\xi_i = \mp\frac{1}{3}$ for $i = 2, 3$, while the Gauss-Lobatto points are placed at $\xi_{\mathrm{GL}\,i} = \mp 0.4472$.
The standard Lagrange element is shown with four Gauss integration points. Their position is given by $\xi_{\mathrm{GP}\,i} = \mp 0.8611$ for $i = 1, 4$ and $\xi_{\mathrm{GP}\,i} = \mp 0.3400$ for $i = 2, 3$ and indicated in Fig. 3.8a with red crosshairs. Even though the difference to the equidistant nodes in the interior is small, none of the Gauss integration points coincide with any of the equidistant nodal positions.
The Gauss-Lobatto-based shape functions, on the other hand, are combined with the Lobatto quadrature scheme, which also uses Gauss-Lobatto points for the integration. The integration points are, therefore, identical to the nodal positions. This is indicated in Fig. 3.8b, where the red crosshairs again show the position of the integration points, now on top of the nodes. Therefore, a strong enforcement of constraints at the nodes leads to a fulfillment of constraints at the integration points. This can be exploited when discretizing a nonlinear manifold as shown in Chapter 5.

Apart from the discretization of manifolds, elements with Gauss-Lobatto-based shape functions are of interest, when using higher-order elements since they do not exhibit Runge's phenomenon [55, 133]. They can, of course, also be combined with other integration rules, however, together with the Lobatto integration they lead to a diagonal mass matrix, which can improve efficiency. More details on the elements with Gauss-Lobatto-based shape functions are given in [133, Chapter 4].

3.4.5. Conclusion

The fundamental basics of rotations were discussed in this chapter and four approaches for a discretization with finite elements were presented. Not all introduced formulations are used in this work. Due to the complexity of some approaches and their consequently high computational cost, they are not feasible options for practical applications. Geodesic finite elements are of limited use in dynamic simulations due to their high computational cost. Also, the projection-based approach for the directors leads to high computational costs and is, thus, limited in practical applications.
Therefore, in Chapter 5 the classical finite element approach and the Gauss-Lobatto-based shape functions are applied to a formulation of the geometrically exact beam using directors. Chapter 6 introduces a formulation of the same beam model in unit quaternions, where the classical finite elements are used as well as the projection-based approach. Additionally, further results for the quaternion formulation with Gauss-Lobatto-based shape functions are given in Appendix C.

4. The geometrically exact beam

Abstract: This chapter covers the theory of the geometrically exact beam formulation. In the first section, we begin with an introduction based on a literature survey, where we discuss the most important aspects of the beam theory and present some applications of the model. The following section discusses the kinematics of the geometrically exact beam in detail. Subsequently, the balance of linear and angular momentum is used to obtain the equations of motion. Furthermore, the kinetic energy is derived and the employed constitutive law is discussed. At the end, a short discussion of the Kirchhoff-Love beam theory is given.

Slender structures play a key role in many mechanical systems from civil and mechanical engineering up to biomechanics and molecular physics [99]. Therefore, there is a high demand for efficient numerical simulations of such slender structures. Mechanical theories describing the behavior of slender structures are referred to as beam or rod theories. The main idea of beam theories is to reduce or limit the physical problem to a one-dimensional theory. Two main goals lie behind the concept of beam theories: In the early days of mechanics, the motivation lay in the possibility to determine general analytic solutions, while today the main motivation is the design of significantly more efficient numerical methods compared to full three-dimensional models to investigate slender geometries.

Depending on the underlying physical problems, beam theories can be classified into either intrinsic, induced, or semi-induced theories [44, 99]. Intrinsic beam theories are directly postulated as one-dimensional theories without derivation from continuum mechanics. However, they still adhere to the essential mechanical principles, such as the balance equations [99]. In contrast to the intrinsic beam formulation, induced beam models are derived from the three-dimensional continuum theory. They describe the motion of the beam under external loads in a three-dimensional setting, using constraints to restrict the movement of the three-dimensional continuum. The kinematic and kinetic relations, as well as stress resultants, can be derived from the three-dimensional theory. Furthermore, even the constitutive relations are related to the continuum theory [99]. In between both approaches lie the semi-induced theories, which obtain the kinematic relations as well as the stress relations from three-dimensional theories, while the material model is postulated [44, 99].

In the non-linear regime, one of the most important beam models is the semi-induced theory of the geometrically exact beam [99]. The naming of that beam model is, unfortunately, ambiguous, which complicates an overview of the topic. In the literature, the same model can be found under various names: (non-linear) Timoshenko beam [46, 65, 67], (special) Cosserat rod [4, 90], Simo-Reissner beam [99] or geometrically exact beam [21, 143]. In this work, we employ the term geometrically exact beam. Equivalently, to the well-known linear theory of the Timoshenko beam, the geometrically exact beam theory, which is utilized in this work, accounts for shear deformations. However, it is not limited to small deformations.

The first approach to a non-linear beam formulation can be attributed to Reissner [112, 113]. A detailed discussion of many aspects of the beam followed shortly after by Antmann [5]. Finally, a full extension of the approach by Reissner to three dimensions was given by Simo in 1985 [126] with numerical examples following in [127–129]. Within the theory proposed by Simo [126] the beam's kinematic and kinetic relation are derived consistently from the continuum theory, which gives reason to the naming of the geometrically exact beam theory. However, the constitutive relations are postulated and are not directly derived from the continuum theory making it a semi-induced theory.

The geometrically exact beam model assumes that each cross-section of the beam remains rigid during deformation. Due to this assumption, which restricts the beam's deformation, the configuration of the beam can be expressed by a position vector and a rotational field. The rotation is expressed through tensors from the special orthogonal group $SO(3)$. The resulting equations are non-linear PDEs, for which an analytical solution can only be obtained in special cases. Thus, to solve problems with arbitrary boundary conditions and for arbitrary geometries, it is usually necessary to employ numerical methods such as the FEM. As discussed in the previous chapter (Section 3.4), a correct discretization approach for the special orthogonal group is not straightforward and cannot be achieved using a classical approach. Therefore, great care has to be taken when discretizing the equation of the beam. The beam formulations found in the literature, thus, mostly differ in the parametrization and discretization of the rotations.

Simo *et al.* [127, 128] proposed a discretization of the incremental rotations. Other approaches, published shortly after, were based on global rotations (e.g. [27, 81]). However, as shown by Crisfield and Jelenić [33] all previously published discretization approaches were neither objective nor path-independent due to their discretization approach of the orthogonal group. To overcome this shortcoming they proposed an objective and path-independent approach based on the underlying structure of the orthogonal group [33], as discussed in Section 3.4.2. Independently, Betsch and Steinmann [21] as well as Romero and Armero [117] proposed an approach, based on the use of directors, which results in an objective and path-independent discretization. Different options for the parametrization and subsequent discretization of the geometrically exact beam are discussed by Romero in [114]. A recent overview over many publications is given by Meier *et al.* in [99].
More recent works deal with the geometrically beam formulation using motion approaches. Sonneville *et al.* [134–136] proposed a new approach, formulating the beam's equation with the help of the special Euclidean group $SE(3)$. This highly sophisticated formulation

enables the design of a locking-free discretized model. A combination of the $SE(3)$ model with warping effects is presented in [119] and in [66] it is used with a Petrov-Galerkin discretization approach. A near-energy conserving integrator for the $SE(3)$ formulation is proposed in [145]. Further, a formulation using dual quaternions is used in [63] to describe the beam's movement in the context of a sliding contact formulation.

Possible application of the geometrically exact beam can be found in the simulation of sliding contacts such as aerial runways [104–107], and for the simulation of twisting wires, for which a general contact condition is discussed in [25]. An embedding of individual beams into a continuum is proposed in [139, 140], where the embedding of a beam in a continuum is used to simulate fiber-reinforced materials. The influence of gravity on the highly nonlinear oscillations of the beam model is investigated in [38]. Till *et al.* [144] use the geometrically exact beam in the context of soft robot models. The beam model is also further extended to incorporate different physical fields. e.g. in [75] the description of the beam is extended to represent electro-mechanical phenomena, which is used in [76] in a context of optimal control.

However, in the present thesis, we focus on a purely mechanical description. This chapter covers the derivation of the mechanical beam model without going into detail about the discretization approach. The theory introduced in this chapter is used in the subsequent chapters, Chapter 5 and Chapter 6, for two different representations of rotations.

As main sources for this chapter, the following sources are used [4, 44, 99, 126]. The chapter is structured as follows. First, we introduce the assumptions of the beam in more detail and describe its kinematics in Section 4.1. We then derive the balance equations in Section 4.2 and the kinetic energy of the beam in Section 4.3. In the next section, Section 4.4, we shortly introduce the postulated constitutive model. Finally, Section 4.5 gives a short discussion of the constrained Kirchhoff-Love beam model.

4.1. Kinematics

We define a beam as a three-dimensional slender continuous body, that extends in one direction far longer than in the other two. The term slender is not defined in a rigorous mathematical manner [4]. A sketch of such a slender geometry is shown in Fig. 4.1. The beam is drawn with a rectangular cross-section. However, the geometry of the body's cross-section can be arbitrary. In the continuum theory, the cross-section deforms, when a physical load is applied. Nevertheless, in the classical beam theory, we assume that the cross-section is rigid and stays plane during the deformation process [4]. Furthermore, during the deformation, the cross-sections do not penetrate each other, which limits the possible bending radius. It cannot be smaller than the radius of the cross-section area [44, Chapter 5].

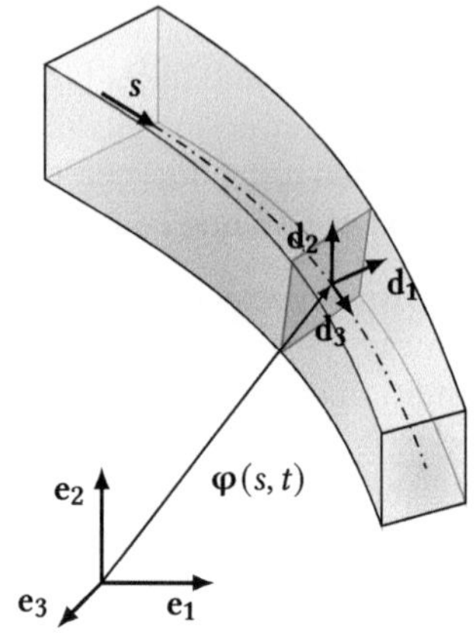

Figure 4.1.: Sketch of a slender body [148].

Along the principle direction of the slender body, a curve is chosen, following the body's largest extension. The curve is parametrized with the arc-length $s \in [0, L]$, where $L \in \mathbb{R} > 0$ is the length of the beam in the undeformed state. The curve can be described by a position vector

$$\boldsymbol{\varphi}(s,t) \in \mathbb{R}^3 . \tag{4.1}$$

At every point of s a cross-section plane is defined, spanned by two vectors $\mathbf{d}_1(s,t) \in \mathbb{R}^3$ and $\mathbf{d}_2(s,t) \in \mathbb{R}^3$ of unit length. The vectors $\mathbf{d}_1(s,t)$ and $\mathbf{d}_2(s,t)$ are orthogonal vectors describing the average orientation of the cross-section at every point of s. As mentioned before, the cross-section itself is assumed to remain rigid. Every point on the cross-section can be described relative to the curve $\boldsymbol{\varphi}(s,t)$ by a vector $\boldsymbol{\theta}(s,t) \in \mathbb{R}^3$ with the basis $\mathbf{d}_1(s,t)$ and $\mathbf{d}_2(s,t)$

$$\boldsymbol{\theta}(s,t) = \theta^\alpha(s)\, \mathbf{d}_\alpha(s,t) . \tag{4.2}$$

As before, Einstein's summation convention applies, and Greek indices α, β take values 1 and 2, while Roman indices i, j, k range from 1 to 3. From here on we omit the arguments (s,t), unless it benefits readability. Together with a third vector $\mathbf{d}_3 \in \mathbb{R}^3$

$$\mathbf{d}_3 = \mathbf{d}_1 \times \mathbf{d}_2 \tag{4.3}$$

the vectors $\mathbf{d}_i$ form an orthonormal frame, often referred to as directors or director frame (see Section 3.3.2). The third director $\mathbf{d}_3$ is normal to the cross-section plane. Note that the numbering of the directors is not uniquely defined in the literature and indeed it may be chosen arbitrarily, as long as the resulting director frame forms a right-handed coordinate system. The notation used in this work is taken from [21]. Since the directors form an orthogonal frame the director frame can be calculated from

$$\mathbf{d}_i = \mathbf{R} \cdot \mathbf{e}_i , \tag{4.4}$$

where $\mathbf{R}(s,t) = \mathbf{d}_i \otimes \mathbf{e}_i$ is a tensor from the special orthogonal group $SO(3)$ describing the orientation of the cross-section. Summarizing the information above, any point of the body shown in Fig. 4.1 can be described by adding the position along the center curve $\boldsymbol{\varphi}$ to the position within the cross-section $\boldsymbol{\theta}$. Therefore, the position vector in the current configuration of every point of the beam results in

$$\mathbf{x}(s,t) = \boldsymbol{\varphi}(s,t) + \boldsymbol{\theta}(s,t) = \boldsymbol{\varphi}(s,t) + \theta^{\alpha}(s)\,\mathbf{d}_{\alpha}(s,t) = \boldsymbol{\varphi}(s,t) + \theta^{\alpha}(s)\,\mathbf{R}(s,t)\cdot\mathbf{e}_{\alpha}\,. \tag{4.5}$$

It follows that the possible configurations of the beam are given by

$$\mathbb{Q} = \{(\boldsymbol{\varphi},\mathbf{R}) : [0,L]\times[0,T] \mapsto \mathbb{R}^3 \times SO(3)\}\,. \tag{4.6}$$

In Fig. 4.2 an arbitrary initial configuration of a beam structure is shown, which is assumed to be stress free. The position of each point on the geometry is given by

$$\mathbf{X}(s) = \boldsymbol{\varphi}_0(s) + \theta^{\alpha}\mathbf{d}_{0\alpha}(s)\,. \tag{4.7}$$

The initial director frame can be computed from

$$\mathbf{d}_{0i}(s) = \mathbf{R}_0(s)\cdot\mathbf{e}_i\,, \tag{4.8}$$

where the orthogonal tensor $\mathbf{R}_0(s)$ defines the initial rotation of the director frame. However, it is usually assumed that in the initial configuration $\mathbf{d}_{03}(s)$ and the normalized tangent vector

$$\mathbf{t}_0 = \frac{\boldsymbol{\varphi}_{0,s}(s)}{\|\boldsymbol{\varphi}_{0,s}(s)\|} \tag{4.9}$$

coincide. In the current configuration, also shown in Fig. 4.2, the direction of the director $\mathbf{d}_3(s,t)$ and the tangent do, in general, no longer coincide due to shear deformation.

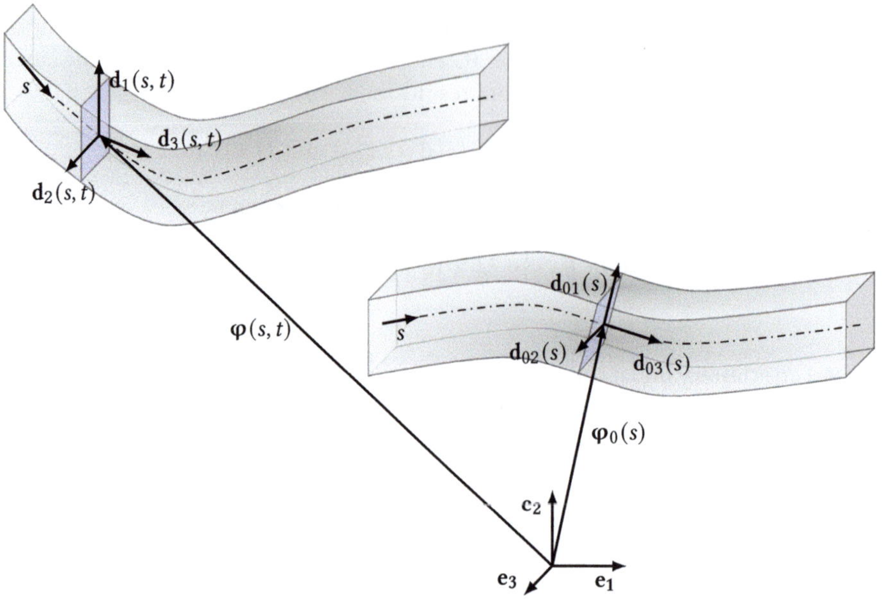

Figure 4.2.: Initial and current configuration of an initially curved beam.

4.1.1. Strain measures

To obtain the first strain measure, the curvature of the beam is examined. The change of rate of the moving director basis $\mathbf{d}_i$ with respect to the parameter s reveals

$$\mathbf{d}_{i,s} = (\mathbf{R} \cdot \mathbf{e}_i)_{,s} = \mathbf{R}_{,s} \cdot \mathbf{e}_i = \mathbf{R}_{,s} \cdot \mathbf{R}^\top \cdot \mathbf{d}_i\,, \tag{4.10}$$

where the abbreviation of the derivative with respect to the parameter s is introduced as $\frac{\partial(\bullet)}{\partial s} = (\bullet)_{,s}$. The curvature $\mathbf{R}_{,s} \cdot \mathbf{R}^\top$ is a skew-symmetric tensor and it describes the change of rate of the rotation in the body fixed frame $\mathbf{d}_i$. The corresponding strain measure can be computed from the curvature through

$$\hat{\mathbf{k}} = \mathbf{R}_{,s} \cdot \mathbf{R}^\top - \mathbf{R} \cdot \mathbf{R}_0^\top \cdot \mathbf{R}_{0,s} \cdot \mathbf{R}^\top\,, \tag{4.11}$$

where $\mathbf{R} \cdot \mathbf{R}_0^\top \cdot \mathbf{R}_{0,s} \cdot \mathbf{R}^\top$ accounts for the curvature of the initial configuration of the beam. The corresponding material description can be obtained by a rotation with $\mathbf{R}$ and is given by [33, 128]

$$\hat{\mathbf{K}} = \mathbf{R}^\top \cdot \mathbf{R}_{,s} - \mathbf{R}_0^\top \cdot \mathbf{R}_{0,s}\,. \tag{4.12}$$

The skew-symmetric tensor $\hat{\mathbf{K}} \in \mathbb{R}^{3\times 3}$ has three independent entries, which can be rewritten into the vector $\mathbf{K}$ with the components

$$K_i = \frac{1}{2}\varepsilon_{ijk}\hat{K}_{kj}\,. \tag{4.13}$$

In the same manner, the spatial axial vector $\mathbf{k}$ can be obtained from the tensor $\hat{\mathbf{k}}$. In the following, we use the vector $\mathbf{K}$ to describe strain measure due to the curvature of the beam. An interpretation of the individual entries can be found in e.g. [4, Chapter VIII]. The two entries K_1 and K_2 account for the bending of the beam, while entry K_3 describes the twist due to torsion.
The shear and dilatation are described through the spatial strain measure [127]

$$\boldsymbol{\gamma} = \boldsymbol{\varphi}_{,s} - \mathbf{d}_3 \,, \tag{4.14}$$

where the entries γ_1 and γ_2 describe the shearing and γ_3 the dilatation of the beam [4, Chapter VIII]. The material measure follows as

$$\boldsymbol{\Gamma} = \mathbf{R}^\top \cdot \boldsymbol{\varphi}_{,s} - \mathbf{e}_3 \,. \tag{4.15}$$

Since the model includes shear deformation it is equivalent to a non-linear formulation of the linear Timoshenko beam. Enforcing $\Gamma_\alpha = 0$ leads to shear stiff beam formulation often referred to as the Kirchhoff-Love beam (cf. Section 4.5).

The three-dimensional strain measures were originally introduced by Simo [126]. They are the work-conjugate strains to the material stress resultants. A detailed discussion of the relations of the strain measures to the theory of deformation of the three-dimensional continuum is given in [8, 27].

4.1.2. Velocity

The velocity vector $\dot{\mathbf{x}}$ is given by the time derivative of the position vector $\mathbf{x}$ in Eq. (4.5) and results in

$$\dot{\mathbf{x}} = \dot{\boldsymbol{\varphi}} + \theta^\alpha \dot{\mathbf{d}}_\alpha \,. \tag{4.16}$$

As before, $\frac{\partial(\bullet)}{\partial t} = \dot{(\bullet)}$ denotes the partial derivative with respect to time. The vector components θ^α do not depend on the time since we assume the cross-sections to be rigid. The change of rate of the directors with respect to time t is computed via

$$\dot{\mathbf{d}}_i = \dot{\mathbf{R}} \cdot \mathbf{e}_i = \dot{\mathbf{R}} \cdot \mathbf{R}^\top \cdot \mathbf{d}_i = \hat{\boldsymbol{\omega}} \cdot \mathbf{d}_i = \boldsymbol{\omega} \times \mathbf{d}_i \,, \tag{4.17}$$

where $\hat{\boldsymbol{\omega}}$ is a skew-symmetric tensor (cf. Section 3.2). The vector

$$\boldsymbol{\omega} = \frac{1}{2} \varepsilon_{ijk} \hat{\omega}_{kj} \mathbf{d}_i \tag{4.18}$$

is the spatial angular velocity as introduced in Section 3.2. The skew-symmetric tensor

$$\mathbf{R}^\top \cdot \dot{\mathbf{R}} = \hat{\boldsymbol{\Omega}} \tag{4.19}$$

incorporates the components of the convective angular velocity vector $\boldsymbol{\Omega}$ by

$$\Omega_i = \frac{1}{2} \varepsilon_{ijk} \hat{\Omega}_{kj} \,. \tag{4.20}$$

4.2. Balance equations

The balance equations of linear and angular momentum are used to derive the equations of motion. For this purpose, we take a segment of the beam $[a, s)$, $0 < a < s < L$, as shown in Fig. 4.3, and apply the respective forces and torques.

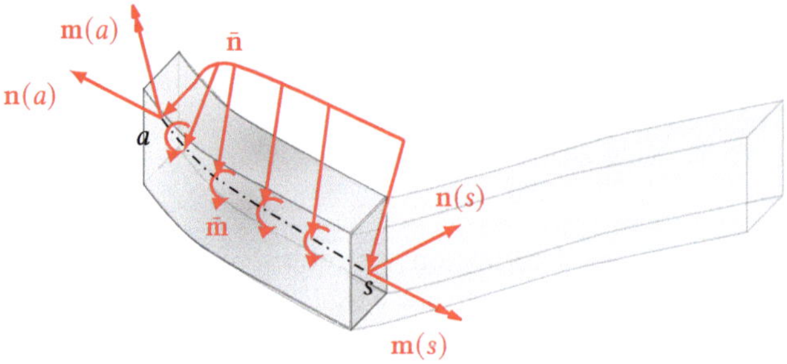

Figure 4.3.: Segment of a beam with applied contact forces and external loads.

The balance equation of linear momentum results in

$$\mathbf{n}(s,t) - \mathbf{n}(a,t) + \int_a^s \bar{\mathbf{n}}(\xi,t)\mathrm{d}\xi = \dot{\mathbf{P}}(s,t)\,, \tag{4.21}$$

where $\mathbf{n}(a,t)$ and $\mathbf{n}(s,t)$ are the resultant contact force at a and s respectively. The distributed external force acting on the curve is denoted by $\bar{\mathbf{n}}(\xi,t)$. The vector $\mathbf{P}$ is the linear momentum. It is given by

$$\begin{aligned}\mathbf{P} &= \int_{\mathcal{V}} \rho_0\,\dot{\mathbf{x}}\mathrm{d}V = \int_a^s \int_{\mathcal{A}} \rho_0 \left(\dot{\boldsymbol{\varphi}} + \theta_\alpha \dot{\mathbf{d}}_\alpha\right) \mathrm{d}A\,\mathrm{d}\xi \\ &= \int_a^s A_\rho \dot{\boldsymbol{\varphi}} + S_\rho^\alpha \cdot \dot{\mathbf{d}}_\alpha\,\mathrm{d}\xi\,.\end{aligned} \tag{4.22}$$

where ρ_0 is the density in the initial configuration. Eq. (4.22) introduces the two inertia terms A_ρ and S_ρ^α. The mass density per unit reference length

$$A_\rho(s) = \int_{\mathcal{A}(s)} \rho_0\,\mathrm{d}A \tag{4.23}$$

is computed by integration of the mass density over the cross-section area $\mathcal{A}(s)$ at s. The static moment is obtained from

$$S_\rho^\alpha(s) = \int_{\mathcal{A}(s)} \rho_0 \theta^\alpha \mathrm{d}A \tag{4.24}$$

for $\alpha = 1, 2$. If the corresponding vector $\mathbf{S}_\rho$ is equal to zero, $\mathbf{S}_\rho = \mathbf{0}$, the curve $\boldsymbol{\varphi}$ coincided with the centerline connecting the centers of gravity of each cross-section [21]. By taking the spatial derivative of Eq. (4.21) we obtain the local form as

$$\mathbf{n}_{,s} + \bar{\mathbf{n}} = \dot{\mathbf{p}}, \tag{4.25}$$

where the linear momentum per unit reference length and its derivative with respect to time are given by

$$\mathbf{p} = A_\rho \dot{\boldsymbol{\varphi}} + S^\alpha_\rho \dot{\mathbf{d}}_\alpha \tag{4.26}$$

and

$$\dot{\mathbf{p}} = A_\rho \ddot{\boldsymbol{\varphi}} + S^\alpha_\rho \ddot{\mathbf{d}}_\alpha \,. \tag{4.27}$$

Analogously, the balance of angular momentum can be derived. In view of Fig. 4.3, the balance of angular momentum at a beam segment can be derived as

$$\begin{aligned} &\mathbf{m}(s,t) - \mathbf{m}(a,t) + (\boldsymbol{\varphi}(s,t) \times \mathbf{n}(s,t)) - (\boldsymbol{\varphi}(a,t) \times \mathbf{n}(a,t)) \\ &\qquad + \int_a^s \boldsymbol{\varphi}(s,t) \times \bar{\mathbf{n}}(s,t) \mathrm{d}s + \int_a^s \bar{\mathbf{m}}(\xi,t) \mathrm{d}\xi = \dot{\mathbf{L}}(s,t) \,, \end{aligned} \tag{4.28}$$

where $\mathbf{m}(s,t)$ and $\mathbf{m}(a,t)$ are the contact moments at a and s. The external distributed torque $\bar{\mathbf{m}}$ acts on the curve. The angular momentum is given through [4, 95]

$$\begin{aligned} \mathbf{L} &= \int_{\mathcal{V}} \rho_0 \, \mathbf{x} \times \dot{\mathbf{x}} \, \mathrm{d}V \\ &= \int_{\mathcal{V}} \rho_0 \left(\boldsymbol{\varphi} + \theta^\alpha \mathbf{d}_\alpha\right) \times \left(\dot{\boldsymbol{\varphi}} + \theta^\beta \dot{\mathbf{d}}_\beta\right) \mathrm{d}V \\ &= \int_a^s \int_{\mathcal{A}} \rho_0 (\boldsymbol{\varphi} \times \dot{\boldsymbol{\varphi}}) + (\theta^\alpha \mathbf{d}_\alpha \times \dot{\boldsymbol{\varphi}}) + (\theta^\beta \boldsymbol{\varphi} \times \dot{\mathbf{d}}_\beta) + \theta^\alpha \theta^\beta \mathbf{d}_\alpha \times \dot{\mathbf{d}}_\beta \, \mathrm{d}A \, \mathrm{d}\xi \,. \end{aligned} \tag{4.29}$$

Using Eq. (4.18) the cross-product in the final term of the integral in Eq. (4.29) can be rewritten as

$$\begin{aligned} \mathbf{d}_\alpha \times \dot{\mathbf{d}}_\beta &= \mathbf{d}_\alpha \times \left(\boldsymbol{\omega} \times \mathbf{d}_\beta\right) \\ &= \left(\mathbf{d}_\alpha \cdot \mathbf{d}_\beta\right) \boldsymbol{\omega} - \left(\mathbf{d}_\beta \otimes \mathbf{d}_\alpha\right) \cdot \boldsymbol{\omega} \,, \end{aligned} \tag{4.30}$$

where the vector identity $\mathbf{u} \times (\mathbf{v} \times \mathbf{w}) = (\mathbf{u} \cdot \mathbf{w})\, \mathbf{v} - (\mathbf{u} \cdot \mathbf{v})\, \mathbf{w}$ has been employed. By aid of Eq. (4.30) it is possible to recast Eq. (4.29) as

$$\mathbf{L} = \int_a^s A_\rho \boldsymbol{\varphi} \times \dot{\boldsymbol{\varphi}} + S^\alpha \left(\mathbf{d}_\alpha \times \dot{\boldsymbol{\varphi}} + \boldsymbol{\varphi} \times (\boldsymbol{\omega} \times \mathbf{d}_\alpha)\right) + J_\rho^{\alpha\beta} (\mathbf{d}_\alpha \otimes \mathbf{d}_\beta) \cdot \boldsymbol{\omega} \, \mathrm{d}\xi \,. \tag{4.31}$$

Therein, $J_\rho^{\alpha\beta}$ are the components of the spatial inertia tensor per unit reference length, which is given by [126]

$$\mathbf{J}_\rho(s) = \int_{\mathcal{A}(s)} \rho_0 \left(\theta^\alpha \theta^\alpha \mathbf{I} - \boldsymbol{\theta} \otimes \boldsymbol{\theta}\right) \mathrm{d}A \,. \tag{4.32}$$

The angular momentum per unit reference length follows from the derivative of Eq. (4.31) as

$$\mathbf{l} = A_\rho \boldsymbol{\varphi} \times \dot{\boldsymbol{\varphi}} + S^\alpha \left(\mathbf{d}_\alpha \times \dot{\boldsymbol{\varphi}} + \boldsymbol{\varphi} \times (\boldsymbol{\omega} \times \mathbf{d}_\alpha)\right) + \mathbf{J}_\rho \cdot \boldsymbol{\omega} \,. \tag{4.33}$$

With the help of Eq. (4.33), the local form of the balance of angular momentum can be written as

$$\mathbf{m}_{,s} + (\boldsymbol{\varphi} \times \mathbf{n})_{,s} + \boldsymbol{\varphi} \times \bar{\mathbf{n}} + \bar{\mathbf{m}} = \dot{\mathbf{l}}, \tag{4.34}$$

where the time derivative of the angular momentum per unit reference length is given by

$$\begin{aligned} \dot{\mathbf{l}} = {} & A_\rho \boldsymbol{\varphi} \times \ddot{\boldsymbol{\varphi}} + S_\rho^\alpha \left(\boldsymbol{\varphi} \times (\dot{\boldsymbol{\omega}} \times \mathbf{d}_\alpha) + \mathbf{d}_\alpha \times \ddot{\boldsymbol{\varphi}}\right) \\ & + \mathbf{J}_\rho \cdot \dot{\boldsymbol{\omega}} + \boldsymbol{\omega} \times \mathbf{J}_\rho \cdot \boldsymbol{\omega} \,. \end{aligned} \tag{4.35}$$

In Eq. (4.35) the following time derivative is used

$$\begin{aligned} \frac{\partial}{\partial t}\left(\mathbf{J}_\rho \cdot \boldsymbol{\omega}\right) &= J_\rho^{\alpha\beta} \dot{\omega}_\beta \mathbf{d}_\alpha + J_\rho^{\alpha\beta} \omega_\beta \dot{\mathbf{d}}_\alpha \\ &= J_\rho^{\alpha\beta} \dot{\omega}_\beta \mathbf{d}_\alpha + \boldsymbol{\omega} \times J_\rho^{\alpha\beta} \omega_\beta \mathbf{d}_\alpha \\ &= \mathbf{J}_\rho \cdot \dot{\boldsymbol{\omega}} + \boldsymbol{\omega} \times \mathbf{J}_\rho \cdot \boldsymbol{\omega} \,. \end{aligned} \tag{4.36}$$

Inserting Eq. (4.25) into Eq. (4.34) and using Eq. (4.35) results in

$$\mathbf{m}_{,s} + \boldsymbol{\varphi}_{,s} \times \mathbf{n} + \bar{\mathbf{m}}(s,t) = S_\rho^\alpha \, \mathbf{d}_\alpha \times \ddot{\boldsymbol{\varphi}} + \mathbf{J}_\rho \cdot \dot{\boldsymbol{\omega}} + \boldsymbol{\omega} \times \mathbf{J}_\rho \cdot \boldsymbol{\omega} \,. \tag{4.37}$$

The two equations, Eq. (4.25) and Eq. (4.37), constitute the equations of motion of the geometrically exact beam [4, Chapter VIII].

4.3. Kinetic energy

The kinetic energy of a continuum body is given by

$$E_{\text{kin}} = \frac{1}{2} \int_{\mathcal{V}} \rho_0(\mathbf{X}) \, \dot{\mathbf{x}}(\mathbf{X}, t) \cdot \dot{\mathbf{x}}(\mathbf{X}, t) \, \mathrm{d}V \,. \tag{4.38}$$

With the help of the inertia terms defined in Section 4.2 in Eq. (4.23), Eq. (4.24) and Eq. (4.32) and the spatial angular velocity from Eq. (4.18) we can write the kinetic energy of the beam as

$$\begin{aligned} E_{\text{kin}} &= \frac{1}{2} \int_{\mathcal{V}} \rho_0(\mathbf{X}) \dot{\mathbf{x}}(\mathbf{X}, t) \cdot \dot{\mathbf{x}}(\mathbf{X}, t) \, \mathrm{d}V \\ &= \frac{1}{2} \int_0^L A_\rho \dot{\boldsymbol{\varphi}} \cdot \dot{\boldsymbol{\varphi}} + 2\dot{\boldsymbol{\varphi}} \cdot (\boldsymbol{\omega} \times \mathbf{d}_\alpha) \, S^\alpha + \boldsymbol{\omega} \cdot \mathbf{J}_\rho \cdot \boldsymbol{\omega} \, \mathrm{d}s \,. \end{aligned} \tag{4.39}$$

In the literature, e.g. in [21, 65, 117, 120], it is often assumed that the curve $\boldsymbol{\varphi}$ coincides with the line of centroids connecting the centers of gravity of each cross-section. Frequently, this is not mentioned explicitly. When both lines fall together $\mathbf{S}_\rho$ is equal to zero [21] as mentioned above. The assumption, that both curves coincide, holds for all examples in this work and we therefore omit the first moment of area $\mathbf{S}_\rho$ from here on.

4.4. Constitutive law

In the following, we discuss the constitutive relation of the geometrically exact beam. In most works covering the geometrically exact beam the beam's behavior is assumed to be hyperelastic, which we also assume. The corresponding stored energy density function per unit length is given by $w(\boldsymbol{\Gamma}, \mathbf{K})$. Taking the partial derivatives of w with respect to the strain measures we obtain the stress resultants

$$\frac{\partial w(\boldsymbol{\Gamma}, \mathbf{K})}{\partial \boldsymbol{\Gamma}} = \mathbf{N}(s,t) \qquad \text{and} \qquad \frac{\partial w(\boldsymbol{\Gamma}, \mathbf{K})}{\partial \mathbf{K}} = \mathbf{M}(s,t)\,, \tag{4.40}$$

where $\mathbf{N}(s,t)$ and $\mathbf{M}(s,t)$ are the resultant force and moment per unit length. The vectors in spatial form can be computed by multiplication from the left with the orthogonal tensor $\mathbf{R}$

$$\mathbf{n} = \mathbf{R} \cdot \mathbf{N} \qquad \text{and} \qquad \mathbf{m} = \mathbf{R} \cdot \mathbf{M}\,. \tag{4.41}$$

Most often a hyperelastic Saint-Vernant-like material is chosen in the literature e.g. in [21, 45, 65, 99, 114, 117, 124, 127, 128] and is employed in this work as well. For a Saint-Vernant-like material, the strain energy function is given by

$$W(\boldsymbol{\Gamma}, \mathbf{K}) = \frac{1}{2} \int_0^L (\boldsymbol{\Gamma} \cdot \mathbf{D}_1 \cdot \boldsymbol{\Gamma} + \mathbf{K} \cdot \mathbf{D}_2 \cdot \mathbf{K})\, \mathrm{d}s\,, \tag{4.42}$$

where the tenors $\mathbf{D}_\alpha$ contain the stiffness parameters of the beam

$$\mathbf{D}_1 = \begin{bmatrix} GA_1 & 0 & 0 \\ 0 & GA_2 & 0 \\ 0 & 0 & EA \end{bmatrix} \qquad \text{and} \qquad \mathbf{D}_2 = \begin{bmatrix} EI_1 & 0 & 0 \\ 0 & EI_2 & 0 \\ 0 & 0 & GI_\mathrm{p} \end{bmatrix}. \tag{4.43}$$

Therein GA_α denotes the shear stiffness along axis $\mathbf{d}_\alpha$ and EA is the axial stiffness. The bending stiffness is given by EI_α with regard to axis $\mathbf{d}_\alpha$ and GI_p is the torsional stiffness of the beam. The directors $\mathbf{d}_1$ and $\mathbf{d}_2$ are, hereby, chosen such that they coincide with the principle axes of the cross-section.
The assumption of rigidity of the cross-sections leads to the necessity that no transversal contraction occurs. In reality, this is only true for materials with a Poisson ratio of $\nu = 0$. However, this effect is usually neglected [99].

Remark 4.1 (Finite strains). *Simo [126] introduced the beam formulation as "a finite strain beam formulation". However, as argued in e.g. [8, 33] the constitutive model, which is applied in the literature (e.g. [45, 65, 99, 114, 117, 124, 127, 128]) is not consistent with finite strains. A detailed discussion is given in [99]. The constitutive relation is indeed only valid for small strains* $\Gamma_i << 1$ *and* $rK_i << 1$*, where* r *is the radius of the cross-section area [99]. The constitutive model can, however, handle finite deformations as long as the strains remain small.*

4.5. The Kirchhoff-Love beam

Apart from beam models with shear deformation, there exist other beam models in the literature with further restrictions on the kinematics e.g. the (inextensible) Kirchhoff-Love beam model [99]. Such models can be viewed as a constrained version of the geometrically exact beam with shear deformations. Similar to the geometrically exact beam, the names for the different constrained formulations are not unique [46, 99].

A Kirchhoff-Love beam model is obtained by introducing additional constraints on the shear strains

$$\Phi_{S\alpha} = \Gamma_\alpha = 0 , \tag{4.44}$$

where the index S indicates constrains on the shear strains. These constraints ensure that $\mathbf{d}_3(s,t)$ does always coincide with the normalized tangent of the centerline

$$\mathbf{t}(s,t) = \frac{\boldsymbol{\varphi}_{,s}(s,t)}{\left\|\boldsymbol{\varphi}_{,s}(s,t)\right\|} \equiv \mathbf{d}_3(s,t) . \tag{4.45}$$

Consequently, the cross-section is always perpendicular to the centerline. Therefore, the model can be viewed as an extension of the linear Bernoulli beam theory.

The shear constraints reduce the configuration manifold from $\mathbb{Q}$ (Eq. (4.6)) to a new set of minimal coordinates

$$\mathbb{Q}_K = \{(\boldsymbol{\varphi}, \psi) : [0, L] \times [0, T] \mapsto \mathbb{R}^3 \times S^1\} . \tag{4.46}$$

Again $\boldsymbol{\varphi}(s,t)$ is the position vector describing the location of the centerline, while $\psi(s,t)$ is an angle describing the torsional movement.

While developing a numerical method for the Kirchhoff-Love beam, it comes naturally to assume, that it is convenient to use the minimal set $\mathbb{Q}_K$. However, the available literature using the minimal set is relatively limited even though many applications exist in various areas (see e.g. [99, 118]). Reasons for this might lie in the complex mathematical framework, which is indeed more complicated than for a beam model with shear deformations.

As mentioned, the director $\mathbf{d}_3$ coincides with the tangent $\mathbf{t}$ in the case of the Kirchhoff-Love model. Thus, without going too much into details, one could assume that the director frame can be determined through the position vector $\boldsymbol{\varphi}$ without problems. Even though $\mathbf{d}_3$ can easily be computed, this does not hold for the other two directors $\mathbf{d}_1$ and $\mathbf{d}_2$, which are needed to describe the configuration of the beam in the entirety. The computation of the directors $\mathbf{d}_1$ and $\mathbf{d}_2$ is equivalent to the computation of vectors, which lie in the tangent space of the unit sphere S^2 spanned by the director $\mathbf{d}_3$. From the Hairy Ball theorem follows that there always exists at least one $\mathbf{d}_3 \in S^2$ for which, the tangent vectors, here $\mathbf{d}_1$ and $\mathbf{d}_2$, in $T_{\mathbf{d}_3}S^2$ are equal to zero [97, 118].
Equivalently, the problem can be comprehended as computing a frame alongside a curve. Multiple options are available for this task. The simplest option is the Serret-Frenet frame [118]. However, it has multiple severe drawbacks, such as it is not defined for a

vanishing curvature and can lead to jumps in the frame, when the curve changes its concavity [97, 118]. So it is not a feasible option, as it would result in unphysical behavior. Another option, the natural or Bishop frame [24], can also not be applied as it is not given explicitly [118].

Thus, usually, an approach using composite rotations is applied (cf. [12, 56, 57, 97, 98]). To the best knowledge of the author, the first objective and path-independent formulation is presented by Meier *et al.* [97] for statics in three dimensions.
In the case of dynamic problems, even less literature exists. In [99] the formulation from [97] is also employed for dynamic examples, but key information about the inertia terms is not presented. An extensive theoretical discussion of the topic can be found in [118] with a detailed FEM model and numerical examples in a subsequent publication [50]. As shown in [118], it is not possible to derive an exact theory for an approach with composite rotations for dynamic problems. Thus, in general, a shear deformable beam formulation as introduced above with additional shear constraints might be more convenient compared to a formulation with minimal coordinates [50, 118].

Additionally, a further constraint can be added, which hinders the elongation of the beam. The inextensibility constrain is given by

$$\Phi_{\mathrm{L}} = \Gamma_3 = 0\,. \tag{4.47}$$

Numerical examples using a constrained formulation for the (inextensible) Kirchhoff-Love beam can be found in e.g. [65].

5. The geometrically exact beam in a director formulation

Abstract: In this chapter, a geometrically exact beam formulation using directors is discussed. In the beginning, a very brief survey of the available literature is given and subsequently, the beam kinematics in this framework are introduced. Furthermore, the kinetic energy is presented with the associated inertia terms. In the following, the continuous weak formulation is given. Subsequently, the approximation approach with finite elements is discussed. It is shown, that the formulation leads to frame-indifferent strain measures. Afterward, the contribution of the inertia terms, the internal forces, and external forces to the weak form is given. Furthermore, orthonormality constraints are introduced, where a weak and a strong enforcement on the control/nodal points are discussed. In the next section the employed time integration scheme, based on the discrete gradient is given. The model's performance and the discretization approach are investigated in static and dynamic examples.

As indicated in the previous chapter, there exists a plethora of publications dealing with the geometrically exact beam. Most of them differ in the way they treat rotations, which often leads to very complex formulations. In this chapter, we investigate a formulation of the geometrically exact beam using a director formulation. As mentioned in the last chapter, the formulation was first proposed by Betsch and Steinmann [21, 22] as well as by Romero and Armero [117] as a frame-indifferent and path-independent alternative to earlier formulations. In [45] this formulation was further advanced to minimize the discretization error introduced by a classical discretization of the director frame. Harsch *et al.* [65] used the formulation in combination with the IGA.

As mentioned in Chapter 2 the Isogeometric Analysis offers many advantages compared to the standard Finite Element Method with Lagrange shape functions. To illustrate some of the benefits we compare both methods in this chapter.

Due to the use of directors, the formulation is relatively simple in its mathematical structure and has only quadratic nonlinearities, which is of great advantage, when constructing energy-momentum conserving time integrators as shown in [22]. Since such conserving integrators have many advantages, especially in regard to numerical stability, we employ such a scheme for the time discretization in this chapter.

We investigate the behavior of standard Lagrange and NURBS shape functions. Additionally, Lagrange shape functions computed based on the Gauss-Lobatto points for the examples in statics are employed. The former two are further used in dynamic examples. The presented approach is not new, but so far to the best knowledge of the author the model was never investigated for higher-order Lagrange elements and elements utilizing the Gauss-Lobatto-based shape functions. Furthermore, a direct comparison of both the FEM and the IGA for this model is so far not presented in the literature. Our aim is to fill this gap here.

The theory of this chapter is based on publications [21, 22, 124], if not mentioned otherwise. The rest of the chapter is structured as follows. In Section 5.1 we introduce the kinematics and strain measures of the geometrically exact beam in the director formulation. The kinetic energy and respective inertia terms are introduced in Section 5.2. The results are used to formulate the weak form of the problem in Section 5.3. Subsequently, Section 5.4, covers the discretization of the weak form, where two options are introduced to ensure the orthonormality of the director frame in the discretized equations. The time integration scheme, which is employed in this chapter, is introduced in Section 5.5. The influence of different discretization approaches and their convergence behavior is investigated in Section 5.6 with several numerical examples.

5.1. Beam kinematics of the director formulation

As shown in the previous chapter, the position of every point on the body of the beam can be described by Eq. (4.5). Most beam formulations parametrize the rotational tensor needed to represent the director frame $\mathbf{d}_i$. But instead of such a parametrization, one can use the directors directly as unknowns as described in Section 3.3.2.
They form a moving frame attached to each cross-section of the beam. Thus, the position vector of each point on the beam can be written without any further parametrization as

$$\mathbf{x}(s,t) = \boldsymbol{\varphi}(s,t) + \theta^{\alpha}(s)\mathbf{d}_{\alpha}(s,t)\,, \tag{5.1}$$

which leads to the following configuration describing the beam's movement

$$\mathbb{Q}_{\mathrm{d}} = \{(\boldsymbol{\varphi}, \mathbf{d}_i) : [0,L] \times [0,T] \mapsto \mathbb{R}^3 \times SO(3)\}\,. \tag{5.2}$$

The configuration manifold is the same as given in Eq. (4.6). Since each director forms the respective column of a rotational tensor, the directors lie in the special orthogonal group $SO(3)$ in the continuous setting. Therefore, the directors are normal and mutual orthogonal in the continuous equations.

5.1.1. Strain measures using directors

As introduced in Chapter 4, the beam is described by two strain measures $\boldsymbol{\Gamma}$ and $\mathbf{K}$. In the director framework they are given by [21]

$$\boldsymbol{\Gamma} = \Gamma_i \mathbf{e}_i\,, \qquad \text{with} \qquad \Gamma_i = \mathbf{d}_i \cdot \boldsymbol{\varphi}_{,s} - \delta_{i3}\,, \tag{5.3}$$

$$\mathbf{K} = K_i \mathbf{e}_i\,, \qquad \text{with} \qquad K_i = \frac{1}{2}\varepsilon_{ijk}\left[\mathbf{d}_k \cdot \mathbf{d}_{j,s} - \mathbf{d}_{0k} \cdot \mathbf{d}_{0j,s}\right]\,. \tag{5.4}$$

5.1.1.1. Frame-indifference of the strain measures

To show that the continuous strain measures are indeed frame-indifferent (or objective) we introduce an arbitrary vector $\mathbf{c}^\# \in \mathbb{R}^3$ describing a finite displacement and an arbitrary rotational tensor $\mathbf{R}^\# \in SO(3)$. Together they describe an arbitrary rigid body movement. With the superimposed rigid body movement, the beam position vector and the director frame are given by

$$\boldsymbol{\varphi}^\# = \mathbf{R} \cdot \boldsymbol{\varphi} + \mathbf{c} \qquad \text{and} \qquad \mathbf{d}_i^\# = \mathbf{R} \cdot \mathbf{d}_i\,. \tag{5.5}$$

To show the frame-indifference of the strain measure $\boldsymbol{\Gamma}$, Eq. (5.5) is inserted into Eq. (5.3) leading to

$$\begin{aligned} \Gamma_i^\# &= \mathbf{d}_i^\# \cdot \boldsymbol{\varphi}_{,s}^\# - \delta_{i3} \\ &= \mathbf{d}_i \cdot \mathbf{R}^\top \cdot \mathbf{R} \cdot \boldsymbol{\varphi}_{,s} - \delta_{i3} \\ &= \mathbf{d}_i \cdot \boldsymbol{\varphi}_{,s} - \delta_{i3} \\ &= \Gamma_i\,. \end{aligned} \tag{5.6}$$

The continuous strain measure $\boldsymbol{\Gamma}$ is, thus, frame-indifferent. In the same manner, the frame-indifference can be shown for the strain measure $\mathbf{K}$ describing the curvature and torsional strain. Inserting Eq. (5.5) into Eq. (5.4) yields

$$\begin{aligned} K_i^\# &= \frac{1}{2}\varepsilon_{ijk}\left[\mathbf{d}_k^\# \cdot \mathbf{d}_{j,s}^\# - \mathbf{d}_{0k}^\# \cdot \mathbf{d}_{0j,s}^\#\right] \\ &= \frac{1}{2}\varepsilon_{ijk}\left[\mathbf{d}_k \cdot \mathbf{R}^\top \cdot \mathbf{R} \cdot \mathbf{d}_{0j,s} - \mathbf{d}_{0k} \cdot \mathbf{R}^\top \cdot \mathbf{R} \cdot \mathbf{d}_{0j,s}\right] \\ &= \frac{1}{2}\varepsilon_{ijk}\left[\mathbf{d}_k \cdot \mathbf{d}_{j,s} - \mathbf{d}_{0k} \cdot \mathbf{d}_{0j,s}\right] \\ &= K_i\,. \end{aligned} \tag{5.7}$$

Therefore, it can be concluded that both continuous strain measures fulfill the requirement to be invariant under superimposed rigid body motions.

5.2. The kinetic energy of the director formulation

We can write the kinetic energy of the geometrically exact beam as

$$E_{\text{kin}} = \frac{1}{2} \int_0^L \int_{\mathcal{A}(s)} \rho_0 \left(\dot{\varphi} + \theta^{\alpha} \dot{\mathbf{d}}_{\alpha} \right) \cdot \left(\dot{\varphi} + \theta^{\beta} \dot{\mathbf{d}}_{\beta} \right) \, \mathrm{d}A \, \mathrm{d}s \,. \tag{5.8}$$

as derived in Section 4.3. The mass density per unit reference length A_ρ, which is the inertia term related to the translational displacements, is not influenced by the choice of the representation of the rotations. It has, therefore, the same form as introduced in Eq. (4.23).
As mentioned in Section 4.3, we assume that $\varphi(s, t)$ is the centerline, connecting the centers of gravity of the cross-sections. From this follows that the coupling terms in the kinetic energy between $\dot{\varphi}$ and $\dot{\mathbf{d}}_i$ are zero ($\mathbf{S}_\rho = \mathbf{0}$).
In contrast to the mass density, the inertia terms with respect to the rotational movement can differ depending on the parametrization. In the case of the director formulation, the Euler tensor describes the inertia of the cross-section with respect to rotational movement [22]. It is given by

$$\mathbf{M}_\rho(s) = \int_{A(s)} \rho_0 \theta^{\alpha} \theta^{\beta} \, \mathbf{d}_{\alpha} \otimes \mathbf{d}_{\beta} \, \mathrm{d}A \,. \tag{5.9}$$

This is in contrast to the inertia tensor (cf. Eq. (4.32)), which is commonly used to represent inertia with respect to rotational movement. The Euler tensor is related to the spatial inertia tensor by

$$\mathbf{J} = M_\rho^{\alpha\alpha} \mathbf{I} - M_\rho^{\alpha\beta} \mathbf{d}_{\alpha} \otimes \mathbf{d}_{\beta} \,. \tag{5.10}$$

Using the Euler tensor in the director formulation the kinetic energy can be written as

$$E_{\text{kin}} = \frac{1}{2} \int_0^L \left[A_\rho \dot{\varphi} \cdot \dot{\varphi} + M_\rho^{\alpha\beta} \dot{\mathbf{d}}_{\alpha} \cdot \dot{\mathbf{d}}_{\beta} \right] \, \mathrm{d}s \,. \tag{5.11}$$

In contrast to the previous chapter $\dot{\mathbf{d}}_i$ are now the velocities of the directors without any further parametrization. Thus, the spatial angular velocity vector $\boldsymbol{\omega}$ in Eq. (4.18) is not used for their computation.
Further, note that the inertia terms do not depend on the configuration but are constant, which is of great advantage for the numerical computation.

5.3. Weak form of the director formulation

The continuous weak form describing the beam's behavior can be written as

$$G_{\text{kin}}(\varphi, \mathbf{d}_i; \delta\varphi, \delta\mathbf{d}_i) + G_{\text{int}}(\varphi, \mathbf{d}_i; \delta\varphi, \delta\mathbf{d}_i) = G_{\text{ext}}(\varphi, \mathbf{d}_i; \delta\varphi, \delta\mathbf{d}_i) \,, \tag{5.12}$$

where the arbitrary admissible variations $(\delta\boldsymbol{\varphi}, \delta\mathbf{d}_i) \in T_{(\varphi,\mathrm{d}_i)}\mathbb{Q}_\mathrm{d}$ lie in the tangent space

$$T_{(\varphi,\mathrm{d}_i)}\mathbb{Q}_\mathrm{d} = \{(\delta\boldsymbol{\varphi}, \delta\mathbf{d}_i) : [0,L] \times [0,T] \mapsto \mathbb{R}^3 \times T_\mathrm{d}SO(3)\}\,, \tag{5.13}$$

where here $T_\mathrm{d}SO(3)$ denotes the tangent space at the rotational tensor $\mathbf{R}$ composed of the directors ($\mathbf{R} = \mathbf{d}_i \otimes \mathbf{e}_i$).

The contribution of the inertia force can be obtained through a variation of the kinetic energy Eq. (5.11), which is given by

$$G_\mathrm{kin}(\mathbf{x}, \delta\mathbf{x}) = \int_0^L A_\rho \delta\boldsymbol{\varphi} \cdot \ddot{\boldsymbol{\varphi}} + M_\rho^{\alpha\beta} \delta\mathbf{d}_\alpha \cdot \ddot{\mathbf{d}}_\beta \, \mathrm{d}s\,. \tag{5.14}$$

Variation of the strain energy function given in Eq. (4.42) reveals the contribution of the internal forces as

$$G_\mathrm{int}(\boldsymbol{\varphi}, \mathbf{d}_i; \delta\boldsymbol{\varphi}, \delta\mathbf{d}_i) = \int_0^L \delta\boldsymbol{\Gamma} \cdot \mathbf{N} + \delta\mathbf{K} \cdot \mathbf{M} \, \mathrm{d}s\,. \tag{5.15}$$

As introduced in the last chapter, we use a Saint-Vernant-like material. It follows that the stress resultants $\mathbf{N}$ and $\mathbf{M}$ are given by $\mathbf{N} = \mathbf{D}_1 \cdot \boldsymbol{\Gamma}$ and $\mathbf{M} = \mathbf{D}_2 \cdot \mathbf{K}$, where the tensors $\mathbf{D}_1$ and $\mathbf{D}_2$ contain the stiffness parameters as given in Section 4.4 by $\mathbf{D}_1 = \mathrm{diag}(GA_1, GA_2, EA)$ and $\mathbf{D}_2 = \mathrm{diag}(EI_1, EI_2, GI_\mathrm{p})$. The variations of the strain measures in Eq. (5.15) result in

$$\delta\boldsymbol{\Gamma} = \delta\Gamma_i \, \mathbf{e}_i \qquad \text{with} \qquad \delta\Gamma_i = \delta\mathbf{d}_i \cdot \boldsymbol{\varphi}_{,s} + \mathbf{d}_i \cdot \delta\boldsymbol{\varphi}_{,s}\,, \tag{5.16}$$

$$\delta\mathbf{K} = \delta K_i \, \mathbf{e}_i \qquad \text{with} \qquad \delta K_i = \frac{1}{2}\varepsilon_{ijk} \left[\delta\mathbf{d}_k \cdot \mathbf{d}_{j,s} + \mathbf{d}_k \cdot \delta\mathbf{d}_{j,s}\right]\,. \tag{5.17}$$

Inserting Eq. (5.16) and Eq. (5.17) into Eq. (5.15) leads to

$$\begin{aligned} G_\mathrm{int}(\boldsymbol{\varphi}, \mathbf{d}_i; \delta\boldsymbol{\varphi}, \delta\mathbf{d}_i) = \int_0^L & \left(\delta\mathbf{d}_i \cdot \boldsymbol{\varphi}_{,s} + \mathbf{d}_i \cdot \delta\boldsymbol{\varphi}_{,s}\right) N_i \\ & + \frac{1}{2}\varepsilon_{ijk} \left(\delta\mathbf{d}_k \cdot \mathbf{d}_{j,s} + \mathbf{d}_k \cdot \delta\mathbf{d}_{j,s}\right) M_i \, \mathrm{d}s\,. \end{aligned} \tag{5.18}$$

The external forces $\bar{\mathbf{n}}$ and torques $\bar{\mathbf{m}}$ lead to the following contribution to the weak form

$$G_\mathrm{ext}(\boldsymbol{\varphi}, \mathbf{d}_i; \delta\boldsymbol{\varphi}, \delta\mathbf{d}_i) = \int_0^L \delta\boldsymbol{\varphi} \cdot \bar{\mathbf{n}} + \frac{1}{2}\delta\mathbf{d}_i \cdot \hat{\bar{\boldsymbol{m}}} \cdot \mathbf{d}_i \, \mathrm{d}s\,, \tag{5.19}$$

where we neglect boundary terms for simplicity. A derivation of the contribution to the weak form of external torques in a director framework in Eq.(5.19) is given in e.g. [17, 124]. The hat operator $(\hat{\bullet})$ denotes again a mapping from the corresponding axial vector to a skew-symmetric tensor.

5.4. Finite element formulation of the director formulation

The approximated solution of the weak form in Eq. (5.12) is obtained using the Finite Element Method or Isogeometric Analysis, respectively. We chose a Bubnov-Galerkin discretization approach. So the approximated displacements and directors as well as their corresponding variations are given by

$$\boldsymbol{\varphi}^h = \sum_{A=1}^{n} N_A \boldsymbol{\varphi}^A \quad \text{and} \quad \delta\boldsymbol{\varphi}^h = \sum_{A=1}^{n} N_A \delta\boldsymbol{\varphi}^A \,, \tag{5.20}$$

$$\mathbf{d}_i^h = \sum_{A=1}^{n} N_A \mathbf{d}_i^A \quad \text{and} \quad \delta\mathbf{d}_i^h = \sum_{A=1}^{n} N_A \delta\mathbf{d}_i^A \,, \tag{5.21}$$

where N_A are Lagrange, standard or Gauss-Lobatto-based, or NURBS shape functions and $\boldsymbol{\varphi}^A$, $\mathbf{d}_i^A$ are the weights of the displacement and directors associated with the A-th node or control point, respectively. The total number of nodes or control points is given by n. As in Chapter 3, we refer from here on to the A-th node or control point as A-th control/nodal point. As mentioned in the introduction, both variables are discretized with a classical finite element approach for the FEM and IGA.

As discussed in detail in Chapter 3 the solution space of the director vectors $\mathbf{d}_i$ is not a linear flat space but the special orthogonal group $SO(3)$. It follows that to achieve optimal convergence behavior for elements of higher-order, a finite element approach should be chosen which conserves the underlying geometry of $SO(3)$. Two options, geodesic and projection-based finite elements, to construct such a conforming finite element space for optimal convergence are discussed in Chapter 3. However, due to their complexity and their high computational costs, an application is not feasible, especially in dynamic problems.
As a possible solution to overcome this shortcoming, we use in this chapter, Gauss-Lobatto-based shape functions beside the standard Lagrange functions.

5.4.1. Discrete strain measures

Employing the discretizations Eq. (5.20) and Eq. (5.21) in Eq. (5.3) and Eq. (5.4) gives the discrete components of the strain measurements

$$\Gamma_i^h = \mathbf{d}_i^h \cdot \boldsymbol{\varphi}_{,s}^h - \delta_{i3} \,, \tag{5.22}$$

$$K_i^h = \frac{1}{2}\varepsilon_{ijk} \left[\mathbf{d}_k^h \cdot \mathbf{d}_{j,s}^h - \mathbf{d}_{0k}^h \cdot \mathbf{d}_{0j,s}^h \right] . \tag{5.23}$$

5.4.1.1. Frame-indifference of the discrete strain measures

As mentioned in Chapter 3 and Chapter 4, all of the early discrete beam formulations suffered from a lack of frame-indifference due to spurious discretization of the rotational variables. Here, it is shown that the discretized director formulation at hand does indeed lead to a frame-indifferent (or objective) formulation when discretized. For this purpose a rigid body motion, described by the vector $\mathbf{c}^{\#} \in \mathbb{R}^3$ and a rotational tensor $\mathbf{R}^{\#} \in SO(3)$, is applied to the control/nodal points analogously to the continuous case in Section 5.1.1.1. The rigid body motion can be applied to the control points due to the affine covariance of NURBS as mentioned in Section 2.3. The values of control/nodal point A under a superimposed rigid body motion are given by

$$\left(\boldsymbol{\varphi}^{A}\right)^{\#} = \mathbf{R} \cdot \boldsymbol{\varphi}^{A} + \mathbf{c} \qquad \text{and} \qquad \left(\mathbf{d}_i^{A}\right)^{\#} = \mathbf{R} \cdot \mathbf{d}_i^{A} \,. \tag{5.24}$$

Inserting Eq. (5.24) into Eq. (5.22) leads to

$$\begin{aligned}
\left(\Gamma_i^h\right)^{\#} &= \left(\mathbf{d}_i^h\right)^{\#} \cdot \left(\boldsymbol{\varphi}^h\right)^{\#}_{,s} - \delta_{i3} \\
&= \sum_{A=1}^{n}\sum_{B=1}^{n} N_A N_{B,s} \left(\mathbf{d}_i^A\right)^{\#} \cdot \left(\boldsymbol{\varphi}^B\right)^{\#} - \delta_{i3} \\
&= \sum_{A=1}^{n}\sum_{B=1}^{n} N_A N_{B,s} \left(\mathbf{d}_i^A \cdot \mathbf{R}^{\top} \cdot \mathbf{R} \cdot \boldsymbol{\varphi}^B + \mathbf{R}^{\top} \cdot \mathbf{d}_i^A \cdot \mathbf{c}\right) - \delta_{i3} \\
&= \mathbf{d}_i^h \cdot \boldsymbol{\varphi}^h_{,s} - \delta_{i3} \\
&= \Gamma_i^h \,,
\end{aligned} \tag{5.25}$$

where relation $\sum_{B=1}^{n} N_{B,s} = 0$, has been used, which follows from the partition of unity of the shape functions. Thus, the discretized strain measure $\boldsymbol{\Gamma}^h$ is frame-indifferent under superimposed rigid body motions. In the same manner, the frame-indifference can be shown for the discrete strain measure $\mathbf{K}^h$ describing the curvature. Inserting Eq. (5.24) into Eq. (5.23) yields

$$\begin{aligned}
\left(K_i^h\right)^{\#} &= \frac{1}{2}\varepsilon_{ijk}\left[\left(\mathbf{d}_k^h\right)^{\#} \cdot \mathbf{d}_{j,s}^{\#} - \left(\mathbf{d}_{0k}^h\right)^{\#} \cdot \left(\mathbf{d}_{0j,s}^h\right)^{\#}\right] \\
&= \frac{1}{2}\varepsilon_{ijk}\left[\mathbf{d}_k^h \cdot \mathbf{R}^{\top} \cdot \mathbf{R} \cdot \mathbf{d}_{j,s}^h - \mathbf{d}_{0k}^h \cdot \mathbf{R}^{\top} \cdot \mathbf{R} \cdot \mathbf{d}_{0j,s}^h\right] \\
&= \frac{1}{2}\varepsilon_{ijk}\left[\mathbf{d}_k^h \cdot \mathbf{d}_{j,s}^h - \mathbf{d}_{0k}^h \cdot \mathbf{d}_{0j,s}^h\right] \\
&= K_i^h \,.
\end{aligned} \tag{5.26}$$

Therefore, both discrete strain measures fulfill the requirement to be frame-indifferent under rigid body motions.

5.4.2. Discrete weak form

Inserting the discretized variables Eq. (5.20) and Eq. (5.21) into the continuous weak form (Eq. (5.12)) reveals the weak form in a finite element framework. However, the discretized directors do not represent the set of minimal coordinates, $\dim(SO(3)) = 3$, but belong to the 9-dimensional space $\mathbb{R}^{3\times 3}$ and are thus redundant. Furthermore, in the discretized equation, we can no longer assume the directors to form an orthonormal frame without enforcing the group constraint of $SO(3)$. Thus, additional Lagrange multipliers are introduced ensuring the normality and mutual orthogonality of the directors and circumventing a redundant formulation. Accordingly, we introduce the discretized weak form

$$\begin{aligned} G_{\mathrm{kin}}(\boldsymbol{\varphi}^h, \mathbf{d}_i^h; \delta\boldsymbol{\varphi}^h, \delta\mathbf{d}_i^h) + G_{\mathrm{int}}(\boldsymbol{\varphi}^h, \mathbf{d}_i^h; \delta\boldsymbol{\varphi}^h, \delta\mathbf{d}_i^h) + G_{\lambda}(\mathbf{d}_i^h, \boldsymbol{\lambda}^h; \delta\mathbf{d}_i^h) \\ = G_{\mathrm{ext}}(\boldsymbol{\varphi}^h, \mathbf{d}_i^h; \delta\boldsymbol{\varphi}^h, \delta\mathbf{d}_i^h)\,, \end{aligned} \tag{5.27}$$

where G_{kin}, G_{int} and G_{ext} are given by Eq. (5.14), Eq. (5.15) and Eq. (5.19), respectively. The discretized Lagrange multipliers $\boldsymbol{\lambda}^h$ ensure the orthonormality of the director frame $\mathbf{d}_i^h$. They contribute with G_{λ} to the weak form. Depending on the employed shape functions, two different formulations of G_{λ} will be used. They are introduced in Section 5.4.2.4.

5.4.2.1. Contribution of the inertia forces

The contribution of the inertia to the discrete weak form can be obtained by inserting Eq. (5.20) and Eq. (5.21) into Eq. (5.14) leading to

$$G_{\mathrm{kin}}(\boldsymbol{\varphi}^h, \mathbf{d}_i^h; \delta\boldsymbol{\varphi}^h, \delta\mathbf{d}_i^h) = \int_0^L A_\rho \delta\boldsymbol{\varphi}^h \cdot \ddot{\boldsymbol{\varphi}}^h + M_\rho^{\alpha\beta}\delta\mathbf{d}_\alpha^h \cdot \ddot{\mathbf{d}}_\beta^h \mathrm{d}s\,. \tag{5.28}$$

Excluding the control/nodal point values the first part of the integral reveals the inertia term associated with the displacements as

$$\mathbf{M}_{AB}^{\varphi} = \int_0^L N_A N_B A_\rho \,\mathbf{I}\,\mathrm{d}s\,, \tag{5.29}$$

and in the same manner from the second term of the integral, the following inertia term associated with the directors is obtained

$$\mathbf{M}_{AB}^{\mathrm{d}} = \int_0^L N_A N_B \begin{bmatrix} M_\rho^{11}\mathbf{I} & M_\rho^{12}\mathbf{I} & \mathbf{0} \\ M_\rho^{21}\mathbf{I} & M_\rho^{22}\mathbf{I} & \mathbf{0} \\ \mathbf{0} & \mathbf{0} & \mathbf{0} \end{bmatrix} \mathrm{d}s\,, \tag{5.30}$$

where $A, B = 1, \ldots, n$. No inertia is associated with the director $\mathbf{d}_3$. The inertia tensor $\mathbf{M}_{AB}^{\mathrm{d}}$ is therefore singular. Using the inertia matrices we can rewrite the contribution to the weak form as

$$G_{\mathrm{kin}}(\boldsymbol{\varphi}^A, \mathbf{d}^A; \delta\boldsymbol{\varphi}^A, \delta\mathbf{d}^A) = \sum_{A=1}^{n}\sum_{B=1}^{n} \begin{bmatrix} \delta\boldsymbol{\varphi}^A \\ \delta\mathbf{d}^A \end{bmatrix}^{\mathrm{T}} \cdot \begin{bmatrix} \mathbf{M}_{AB}^{\varphi} & \mathbf{0} \\ \mathbf{0} & \mathbf{M}_{AB}^{\mathrm{d}} \end{bmatrix} \cdot \begin{bmatrix} \ddot{\boldsymbol{\varphi}}^B \\ \ddot{\mathbf{d}}^B \end{bmatrix}, \tag{5.31}$$

where the vector $\mathbf{d}^A \in \mathbb{R}^9$ contains the control/nodal point values of all three directors $\mathbf{d}_i^A$ of the A-th control/nodal point in an ordered manner

$$\mathbf{d}^A = \begin{bmatrix} \mathbf{d}_1^A & \mathbf{d}_2^A & \mathbf{d}_3^A \end{bmatrix}^\top . \tag{5.32}$$

5.4.2.2. Contribution of the internal forces

The internal forces contribute to the weak form with

$$G_{\text{int}}(\boldsymbol{\varphi}^h, \mathbf{d}_i^h; \delta\boldsymbol{\varphi}^h, \delta\mathbf{d}_i^h) = \int_0^L \delta\boldsymbol{\Gamma}^h \cdot \mathbf{N}^h + \delta\mathbf{K}^h \cdot \mathbf{M}^h \mathrm{ds}, \tag{5.33}$$

where $\mathbf{N}^h = \mathbf{D}_1 \cdot \boldsymbol{\Gamma}^h$ and $\mathbf{M}^h = \mathbf{D}_2 \cdot \mathbf{K}^h$ are the discretized stress resultants. Using the discrete strain measures Eq. (5.22) and Eq. (5.23) allows to recast Eq. (5.33) as

$$\begin{aligned} G_{\text{int}}(\boldsymbol{\varphi}^h, \mathbf{d}_i^h; \delta\boldsymbol{\varphi}^h, \delta\mathbf{d}_i^h) = \int_0^L & \left(\delta\mathbf{d}_i^h \cdot \boldsymbol{\varphi}_{,s}^h + \mathbf{d}_i^h \cdot \delta\boldsymbol{\varphi}_{,s}^h\right) N_i^h \\ & + \frac{1}{2}\varepsilon_{ijk}\left(\delta\mathbf{d}_k^h \cdot \mathbf{d}_{j,s}^h + \mathbf{d}_k^h \cdot \delta\mathbf{d}_{j,s}^h\right) M_i^h \mathrm{ds}, \end{aligned} \tag{5.34}$$

which we rewrite into

$$G_{\text{int}}(\boldsymbol{\varphi}^A, \mathbf{d}^A; \delta\boldsymbol{\varphi}^A, \delta\mathbf{d}^A) = \sum_{A=1}^{n} \begin{bmatrix} \delta\boldsymbol{\varphi}^A \\ \delta\mathbf{d}^A \end{bmatrix}^\top \cdot \int_0^L \begin{bmatrix} \mathbf{B}_{\varphi\varphi}^A & \mathbf{0} \\ \mathbf{B}_{\mathrm{d}\varphi}^A & \mathbf{B}_{\mathrm{dd}}^A \end{bmatrix} \cdot \begin{bmatrix} \mathbf{N}^h \\ \mathbf{M}^h \end{bmatrix} \mathrm{ds}. \tag{5.35}$$

The nodal operator matrices are given by

$$\begin{aligned} \mathbf{B}_{\varphi\varphi}^A &= N_{A,s} \begin{bmatrix} \mathbf{d}_1^h & \mathbf{d}_2^h & \mathbf{d}_3^h \end{bmatrix}, \\ \mathbf{B}_{\mathrm{d}\varphi}^A &= N_A \begin{bmatrix} \boldsymbol{\varphi}_{,s}^h & \mathbf{0} & \mathbf{0} \\ \mathbf{0} & \boldsymbol{\varphi}_{,s}^h & \mathbf{0} \\ \mathbf{0} & \mathbf{0} & \boldsymbol{\varphi}_{,s}^h \end{bmatrix}, \\ \mathbf{B}_{\mathrm{dd}}^A &= \frac{1}{2} N_A \begin{bmatrix} \mathbf{0} & -\mathbf{d}_{3,s}^h & \mathbf{d}_{2,s}^h \\ \mathbf{d}_{3,s}^h & \mathbf{0} & -\mathbf{d}_{1,s}^h \\ -\mathbf{d}_{2,s}^h & \mathbf{d}_{1,s}^h & \mathbf{0} \end{bmatrix} + \frac{1}{2} N_{A,s} \begin{bmatrix} \mathbf{0} & \mathbf{d}_3^h & -\mathbf{d}_2^h \\ -\mathbf{d}_3^h & \mathbf{0} & \mathbf{d}_1^h \\ \mathbf{d}_2^h & -\mathbf{d}_1^h & \mathbf{0} \end{bmatrix}. \end{aligned} \tag{5.36}$$

Alternatively Eq. (5.35) can be written as

$$G_{\text{int}}(\boldsymbol{\varphi}^A, \mathbf{d}^A; \delta\boldsymbol{\varphi}^A, \delta\mathbf{d}^A) = \sum_{A=1}^{n} \delta\boldsymbol{\varphi}^A \cdot \mathbf{f}_{\text{int}}^{\varphi,A} + \delta\mathbf{d}^A \cdot \mathbf{f}_{\text{int}}^{\mathrm{d},A}, \tag{5.37}$$

where the following internal nodal forces are used

$$\begin{aligned} \mathbf{f}_{\text{int}}^{\varphi,A} &= \int_0^L \mathbf{B}_{\varphi\varphi}^A \cdot \mathbf{N}^h \mathrm{ds}, \\ \mathbf{f}_{\text{int}}^{\mathrm{d},A} &= \int_0^L \mathbf{B}_{\mathrm{d}\varphi}^A \cdot \mathbf{N}^h + \mathbf{B}_{\mathrm{dd}}^A \cdot \mathbf{M}^h \mathrm{ds}. \end{aligned} \tag{5.38}$$

5.4.2.3. Contribution of external loading

Using Eq. (5.19) together with Eq. (5.20) and Eq. (5.21) the contribution of the external loads towards the the discretized weak form is given by

$$G_{\text{ext}}(\boldsymbol{\varphi}^h, \mathbf{d}_i^h; \delta\boldsymbol{\varphi}^h, \delta\mathbf{d}_i^h) = \int_0^L \delta\boldsymbol{\varphi}^h \cdot \bar{\mathbf{n}} + \frac{1}{2}\delta\mathbf{d}_i^h \cdot \hat{\bar{\boldsymbol{m}}} \cdot \mathbf{d}_i^h \mathrm{d}s\,, \tag{5.39}$$

which can be recast as

$$G_{\text{ext}}(\boldsymbol{\varphi}^A, \mathbf{d}_i^A; \delta\boldsymbol{\varphi}^A, \delta\mathbf{d}_i^A) = \sum_{A=1}^{n} \delta\boldsymbol{\varphi}^A \cdot \mathbf{f}_{\text{ext}}^{\varphi,A} + \delta\mathbf{d}_i^A \cdot \mathbf{f}_{\text{ext}}^{\mathrm{d},A} \mathrm{d}s \tag{5.40}$$

using the control/nodal point forces given by

$$\begin{aligned} \mathbf{f}_{\text{ext}}^{\varphi,A} &= \int_0^L N_A \bar{\mathbf{n}} \mathrm{d}s\,, \\ \mathbf{f}_{\text{ext}}^{\mathrm{d},A} &= \int_0^L \frac{1}{2} N_A \hat{\bar{\boldsymbol{m}}} \cdot \mathbf{d}_i^h \mathrm{d}s\,. \end{aligned} \tag{5.41}$$

5.4.2.4. Contribution of the orthonormality constraints

As mentioned in Section 5.4.2, the redundancy of the discrete director formulation can be reduced using Lagrange multipliers, where the normality and mutual orthogonality of the directors are enforced using the following constraint equation

$$\bar{g}_{ij}(\mathbf{d}_i) = \frac{1}{2}\left(\mathbf{d}_i \cdot \mathbf{d}_j - \delta_{ij}\right) = 0\,, \tag{5.42}$$

which results in six independent constraints due to the symmetry of δ_{ij}. In the following, it is more convenient to write the constraints in a vector format

$$\mathbf{g}(\mathbf{d}_i) = \frac{1}{2}\begin{bmatrix} \mathbf{d}_1 \cdot \mathbf{d}_1 - 1 \\ \mathbf{d}_2 \cdot \mathbf{d}_2 - 1 \\ \mathbf{d}_3 \cdot \mathbf{d}_3 - 1 \\ \mathbf{d}_1 \cdot \mathbf{d}_2 \\ \mathbf{d}_1 \cdot \mathbf{d}_3 \\ \mathbf{d}_2 \cdot \mathbf{d}_3 \end{bmatrix} = \mathbf{0}\,, \tag{5.43}$$

where the first three entries ensure the normality and the last three entries the mutual orthogonality. In the following, we discuss two options to enforce the constraints.

Weak enforcement First, we introduce a weak enforcement, which is used in [65] with NURBS shape functions and the director formulation. For this purpose, the Lagrange multipliers are discretized using a Bubnov-Galerkin approach

$$\boldsymbol{\lambda}^h = \sum_{A=1}^{n} N_A^{\lambda} \boldsymbol{\lambda}^A \quad \text{and} \quad \delta\boldsymbol{\lambda}^h = \sum_{A=1}^{n} N_A^{\lambda} \delta\boldsymbol{\lambda}^A, \tag{5.44}$$

where $\boldsymbol{\lambda}^A \in \mathbb{R}^6$ are the control/nodal point values of the Lagrange multipliers. This leads to an integral or weak enforcement of the orthonormality constraints, which results in the following contribution to the weak form

$$G_\lambda(\mathbf{d}_i^h, \boldsymbol{\lambda}^h; \delta\mathbf{d}_j^h) = \int_0^L \delta\mathbf{g}(\mathbf{d}_i^h) \cdot \boldsymbol{\lambda}^h \mathrm{d}s = \int_0^L \delta\mathbf{d}_j^h \cdot \mathbf{G}(\mathbf{d}_i^h)^\top \cdot \boldsymbol{\lambda}^h \mathrm{d}s, \tag{5.45}$$

where

$$\mathbf{G}(\mathbf{d}_i) = \nabla_{\mathrm{d}} \mathbf{g}(\mathbf{d}_i) \tag{5.46}$$

is the gradient of the constraints (Eq. (5.43)) with respect to the directors symbolized with the nabla operator ∇. Additionally, the orthonormality of the directors is ensured by

$$\int_0^L \delta\boldsymbol{\lambda}^h \cdot \mathbf{g}(\mathbf{d}_i^h) \mathrm{d}s = 0. \tag{5.47}$$

Using Eq. (5.44) the discrete constraint functions are given by

$$\mathbf{g}^A = \int_0^L N_A^{\lambda} \mathbf{g}(\mathbf{d}_i^h) \mathrm{d}s \tag{5.48}$$

with $A = 1, \ldots, n$. Rewriting the contribution towards the weak form in terms of the discrete constraint forces leads to

$$G_\lambda(\mathbf{d}_i^A, \boldsymbol{\lambda}^n; \delta\mathbf{d}_j^A) = \sum_{i=1}^{A} \delta\mathbf{d}_j^A \cdot \mathbf{f}_\lambda^A, \tag{5.49}$$

where the control/nodal point forces are given by

$$\mathbf{f}_\lambda^A = \int_0^L N_A^{\lambda} \mathbf{G}(\mathbf{d}_i^h)^\top \cdot \boldsymbol{\lambda}^h \mathrm{d}s. \tag{5.50}$$

Strong enforcement As the second option to enforce the orthonormality constraints, a point-wise or strong enforcement is introduced. A point-wise enforcement on the control/nodal points leads to

$$\mathbf{g}^A = \mathbf{g}(\mathbf{d}^A) = \frac{1}{2} \begin{bmatrix} \mathbf{d}_1^A \cdot \mathbf{d}_1^A - 1 \\ \mathbf{d}_2^A \cdot \mathbf{d}_2^A - 1 \\ \mathbf{d}_3^A \cdot \mathbf{d}_3^A - 1 \\ \mathbf{d}_1^A \cdot \mathbf{d}_2^A \\ \mathbf{d}_1^A \cdot \mathbf{d}_3^A \\ \mathbf{d}_2^A \cdot \mathbf{d}_3^A \end{bmatrix} = \mathbf{0} \tag{5.51}$$

for the A-th control/nodal point as discrete constraint functions. The control/nodal point forces due to the constraints, employed in Eq. (5.49) in the contribution to the weak form, follow from

$$\mathbf{f}_{\lambda}^{A} = \mathbf{G}(\mathbf{d}_i^A)^{\top} \cdot \boldsymbol{\lambda}^A . \tag{5.52}$$

for the strong enforcement. The strong enforcement is chosen in combination with a discretization of the other variables with linear Lagrange shape functions in many publications e.g. in [21, 22, 45, 117, 124].

5.4.3. Differential-algebraic system of equations

The discretization of the weak form with the Finite Element Method leads to the following semi-discrete system of differential-algebraic equations (DAEs)

$$\sum_{A=1}^{n} \begin{bmatrix} \delta\boldsymbol{\varphi}^A \\ \delta\mathbf{d}^A \\ \delta\boldsymbol{\lambda}^A \end{bmatrix}^{\top} \cdot \left(\sum_{B=1}^{n} \begin{bmatrix} \mathbf{M}_{AB}^{\varphi}\, \ddot{\boldsymbol{\varphi}}^B \\ \mathbf{M}_{AB}^{\mathrm{d}}\, \ddot{\mathbf{d}}^B \\ \mathbf{0} \end{bmatrix} + \begin{bmatrix} \mathbf{f}_{\mathrm{int}}^{\varphi,A} \\ \mathbf{f}_{\mathrm{int}}^{\mathrm{d},A} \\ \mathbf{0} \end{bmatrix} + \begin{bmatrix} \mathbf{0} \\ \mathbf{f}_{\lambda}^{\mathrm{d},A} \\ \mathbf{g}^A \end{bmatrix} - \begin{bmatrix} \mathbf{f}_{\mathrm{ext}}^{\varphi,A} \\ \mathbf{f}_{\mathrm{ext}}^{\mathrm{d},A} \\ \mathbf{0} \end{bmatrix} \right) = 0 \tag{5.53}$$

as the variations of the control/nodal points $(\delta\boldsymbol{\varphi}^A, \delta\mathbf{d}^A, \delta\boldsymbol{\lambda}^A) \in \mathbb{R}^3 \times \mathbb{R}^9 \times \mathbb{R}^6$ are arbitrary for all $A \in 1, \ldots, n$. One can choose between a weak or strong enforcement of the constraints by choosing $\mathbf{f}_{\lambda}^{\mathrm{d},A}$ and $\mathbf{g}^A$ accordingly.
The system of DAEs can be rewritten into a more convenient format exploiting the arbitrariness of the variational control/nodal point values $(\delta\boldsymbol{\varphi}^A, \delta\mathbf{d}^A, \delta\boldsymbol{\lambda}^A)$. For this purpose, all the control/nodal point values for displacements and the directors are assembled in a vector $\mathbf{q} \in \mathbb{R}^{12n}$ and the control/nodal point values of the Lagrange multipliers are gathered in $\boldsymbol{\lambda} \in \mathbb{R}^{6n}$. We obtain the following assembled time-continuous system of DAEs

$$\begin{aligned} \mathbf{M} \cdot \ddot{\mathbf{q}}(t) + \mathbf{f}_{\mathrm{int}}(t) + \mathbf{G}_{\mathrm{g}}(t)^{\top} \cdot \boldsymbol{\lambda}(t) &= \mathbf{f}_{\mathrm{ext}}(t)\,, \\ \boldsymbol{\Phi}\left(\mathbf{q}(t)\right) &= \mathbf{0}\,, \end{aligned} \tag{5.54}$$

where $\mathbf{M}$ contains the assembled inertia matrices, $\mathbf{f}_{\mathrm{int}}(t)$ all the control/nodal point forces due to the internal work, $\mathbf{G}_{\mathrm{g}}$ the assembled gradients of the constraints and $\boldsymbol{\Phi}$ the constraint functions.

Remark 5.1 (Strong enforcement of constraints in the IGA). *Various approaches were investigated in the course of this work to enforce the orthonormality constraints point-wise in combination with B-Splines and NURBS: An enforcement at points computed via a closest-point projection, equally spaced points in the physical domain, as well as in the parameter space, and the enforcement on the control points, themselves were explored. Our numerical experiments show that all these approaches for strong enforcement do not result in a reliable method and do not lead to convergence of the employed Newton's method for arbitrary meshes.*

5.5. Discretization in time

Even though the topic of time integration schemes has a long history, it is still part of ongoing research. Especially, efficient and robust integrators for nonlinear differential-algebraic equations are still a challenging task. Conserving integrators, which preserve as much as possible of the underlying structure, are of advantage here, as described in many publications (e.g. [22, 23, 130, 131]). We use an energy-momentum scheme based on the discrete gradient, which conserves the energy, the linear and angular momentum of the fully discretized system. The discrete gradient was first proposed by Gonzalez in [53] and expanded to systems with holonomic constraints in [54].

For the integration in time we divide the time domain $\mathcal{I}$ of interest into $k \in \mathbb{N}$ subintervals

$$\mathcal{I} = [0, T] = \bigcup_{n=0}^{k-1} [t_n, t_{n+1}] \,. \tag{5.55}$$

The length of the n-th time interval is given by $\Delta t_n = t_{n+1} - t_n$. Using the energy-momentum scheme based on the discrete gradient the fully discretized equations for the geometrically exact beam are given by

$$\begin{aligned} \mathbf{M} \cdot \mathbf{a}_{n+\frac{1}{2}} + \bar{\mathbf{f}}_{\text{int}} \left(\mathbf{q}_n, \mathbf{q}_{n+1}\right) + \mathbf{G}_{\text{g}} \left(\mathbf{q}_{n+\frac{1}{2}}\right)^{\mathsf{T}} \cdot \boldsymbol{\lambda}_{n+1} &= \mathbf{f}_{\text{ext}}(t_{n+\frac{1}{2}}) \,, \\ \boldsymbol{\Phi}(\mathbf{q}_{n+1}) &= \mathbf{0} \end{aligned} \tag{5.56}$$

with

$$\mathbf{q}_{n+\frac{1}{2}} = \frac{1}{2} \left(\mathbf{q}_n + \mathbf{q}_{n+1}\right) \,, \tag{5.57}$$

$$\mathbf{v}_n = \frac{2}{\Delta t_n} (\mathbf{q}_n - \mathbf{q}_{n-1}) - \mathbf{v}_{n-1} \,, \tag{5.58}$$

$$\mathbf{a}_{n+\frac{1}{2}} = \frac{2}{\Delta t_n^2} \left(\mathbf{q}_{n+1} - \mathbf{q}_n\right) - \frac{2}{\Delta t_n} \mathbf{v}_n \,. \tag{5.59}$$

The control/nodal point values evaluated at time t_n are $\mathbf{q}_n$, and respectively, $\mathbf{q}_{n+1}$ are the control/nodal point values at time t_{n+1}, analogous for $\boldsymbol{\lambda}_{n+1}$and $\mathbf{v}_n$, where $\mathbf{v}$ are the nodal velocities.
The conservation properties of the integration scheme in Eq. (5.56) is, of course, only given if no external loads are applied ($\mathbf{f}_{\text{ext}} = \mathbf{0}$).

The operator $(\bar{\bullet})$ used for $\bar{\mathbf{f}}_{\text{int}}$ indicates the use of the discrete gradient $\bar{\boldsymbol{\nabla}}$ to derive the internal forces from the internal potential. The discrete gradient $\bar{\boldsymbol{\nabla}} V$ of a function V introduced in [53] is given by

$$\bar{\boldsymbol{\nabla}} V(\mathbf{q}_n, \mathbf{q}_{n+1}) = \boldsymbol{\nabla} V(\mathbf{q}_{n+\frac{1}{2}}) + \frac{V(\mathbf{q}_{n+1}) + V(\mathbf{q}_n) - \boldsymbol{\nabla} V(\mathbf{q}_{n+\frac{1}{2}}) \cdot (\mathbf{q}_{n+1} - \mathbf{q}_n)}{\left\|(\mathbf{q}_{n+1} - \mathbf{q}_n)\right\|^2} \cdot (\mathbf{q}_{n+1} - \mathbf{q}_n) \,. \tag{5.60}$$

According to Gonzalez [53] the discrete gradient has to fulfill the following two properties: the directionality condition

$$\bar{\nabla} V\left(\mathbf{q}_n, \mathbf{q}_{n+1}\right) \cdot \left(\mathbf{q}_{n+1} - \mathbf{q}_n\right) = V(\mathbf{q}_{n+1}) - V\left(\mathbf{q}_n\right), \tag{5.61}$$

and the consistency condition

$$\bar{\nabla} V\left(\mathbf{q}_n, \mathbf{q}_{n+1}\right) = \bar{\nabla} V\left(\frac{\mathbf{q}_n + \mathbf{q}_{n+1}}{2}\right) + \mathcal{O}\left(\|\mathbf{q}_{n+1} - \mathbf{q}_n\|\right). \tag{5.62}$$

For a quadratic function, the discrete gradient coincides with the mid-point evaluation [23], which is used in Eq. (5.56) to evaluate the gradient of the constraints. In the case of an energy function V given by a Saint-Vernant material and a strain measure $\mathbf{E}$, which depends quadratically on the control/nodal points, the discrete gradient coincides with the use of the average of the strains

$$\mathbf{E}_{n+\frac{1}{2}}(\mathbf{q}_n, \mathbf{q}_{n+1}) = \frac{1}{2}\left(\mathbf{E}(\mathbf{q}_n) + \mathbf{E}(\mathbf{q}_{n+1})\right) \tag{5.63}$$

such that

$$\bar{\nabla} V\left(\mathbf{q}_n, \mathbf{q}_{n+1}\right) = \nabla V\left(\mathbf{E}_{n+\frac{1}{2}}\right). \tag{5.64}$$

Simo and Tarnow [130] discovered the structure-conserving properties of the average of the strains in Eq. (5.63) for systems with a Saint-Vernant material and quadratic strain measures.
In the case of the director formulation, all strain measures are quadratic and we use a Saint-Vernant-type material behavior. So $\bar{\mathbf{f}}_{int}$ is evaluated using the average of the strain measures $\boldsymbol{\Gamma}_{n+\frac{1}{2}}$ and $\mathbf{K}_{n+\frac{1}{2}}$, which are computed according to Eq. (5.63).
A detailed discussion on the topic of the discrete gradient, its uniqueness, and physical interpretation is given in [116].

The nonlinear equation Eq. (5.56) is solved using the Newton method. The update in the Netwon algorithm is performed additively. Due to the application of the discrete gradient, total energy as well as linear and angular momentum, are conserved within the accuracy range of the tolerance of the Newton method.

5.5.1. Elimination of the Lagrange multipliers and constraints

The strong enforcement of the orthonormality constraints allows the application of an algorithm to eliminate the discrete Lagrange multipliers as well as the constraint equations. This approach, which is referred to as the discrete null space method, is discussed in detail [18, 19] including time integration with the discrete gradient. In the context of the geometrically exact beam formulation based on directors the null-space method is covered in [92].
To achieve the elimination of the Lagrange multipliers, the control/nodal point velocities are expressed by $\dot{\mathbf{q}}^A = \mathbf{P}^A \cdot \mathbf{t}^A$, where $\mathbf{t}^A \in \mathbb{R}^6$ are the generalized velocities

$\mathbf{t}^A = \begin{bmatrix} \dot{\boldsymbol{\varphi}}^A & \boldsymbol{\omega}^A \end{bmatrix}^\top$ of control/nodal point A. The null space matrix $\mathbf{P}^A$ for the A-th control/nodal point is given by

$$\mathbf{P}^A = \begin{bmatrix} \mathbf{I} & \mathbf{0} & \mathbf{0} & \mathbf{0} \\ \mathbf{0} & -\hat{\mathbf{d}}_1^A & -\hat{\mathbf{d}}_2^A & -\hat{\mathbf{d}}_3^A \end{bmatrix}^\top . \tag{5.65}$$

The null space matrices $\mathbf{P}^A$ of each control/nodal point are assembled in a global null space matrix $\mathbf{P}$. By premultiplying Eq. $(5.56)_1$ from the left with the global discrete null space matrix $\mathbf{P}(\mathbf{q}_{n+\frac{1}{2}})$ we obtain the size-reduced system

$$\mathbf{P}(\mathbf{q}_{n+\frac{1}{2}})^\top \cdot \left(\mathbf{M}\mathbf{a}_{n+\frac{1}{2}} + \bar{\mathbf{f}}_{\text{int}}\left(\mathbf{q}_n, \mathbf{q}_{n+1}\right)\right) = \mathbf{P}(\mathbf{q}_{n+\frac{1}{2}})^\top \cdot \mathbf{f}_{\text{ext}}(t_{n+\frac{1}{2}}) . \tag{5.66}$$

By construction $\mathbf{P}(\mathbf{q}_{n+\frac{1}{2}})$ lies in the null space of the gradient of the constrains $\mathbf{G}_{\text{g}}(\mathbf{q}_{n+\frac{1}{2}})$, thus, they are orthogonal ($\mathbf{G}_{\text{g}} \cdot \mathbf{P} = \mathbf{0}$). The constraints in Eq. $(5.56)_2$ still need to be fulfilled, which can be achieved through an appropriate update procedure for the directors, which conserves the unit length and their mutual orthogonality. Such an update procedure is given by

$$\left(\mathbf{d}_i^A\right)_{n+1} = \mathbf{R}(\boldsymbol{\theta}^A) \cdot \left(\mathbf{d}_i^A\right)_n , \tag{5.67}$$

where $\boldsymbol{\theta}^A \in \mathbb{R}^3$ are incremental rotations, which rotate the director frame in between the individual load or time increments. These incremental rotations are the new unknowns. The rotational tensor in Eq. (5.67) can be computed using the exponential map of $SO(3)$ via

$$\mathbf{R}(\boldsymbol{\theta}^A) = \exp_{SO(3)}(\boldsymbol{\theta}^A) = \mathbf{I} + \frac{\sin\left(\|\boldsymbol{\theta}^A\|\right)}{\|\boldsymbol{\theta}^A\|}\hat{\boldsymbol{\theta}}^A + \frac{1}{2}\left(\frac{\sin\left(\frac{1}{2}\|\boldsymbol{\theta}^A\|\right)}{\frac{1}{2}\|\boldsymbol{\theta}^A\|}\right)^2 \left(\hat{\boldsymbol{\theta}}^A\right)^2 , \tag{5.68}$$

where $(\hat{\bullet})$ denotes again a mapping from $\mathbb{R}^3 \mapsto \mathfrak{so}(3)$. The update given by Eq. (5.67) ensures the orthonormality of the directors at every control/nodal point. During each time or load step the incremental rotations $\boldsymbol{\theta}^A$ are updated additively.

5.6. Numerical validation of the director formulation

Abbreviations for the different elements used in this chapter are given in Tab. 5.1. The Lagrange shape functions are combined with a strong enforcement of the orthonormality constraints, while when using NURBS shape functions the constraints are enforced in an integral or weak sense. For example, the abbreviation NCW3 describes an element with NURBS shape functions of order 3 with weak enforcement of the orthonormality constraints. LCS2 implies an element using the standard Lagrange basis of order 2 with strong enforcement of the constraints. An element using the Gauss-Lobatto-based shape functions with a strong enforcement is abbreviated by GCS1.

Table 5.1.: Abbreviations for the employed elements.

shape functions	constraints	order	abbreviation
		1	NCW1
NURBS	weak	2	NCW2
		3	NCW3
		1	LCS1
Lagrange	strong	2	LCS2
		3	LCS3
		1	GCS1
Gauss-Lobatto-based	strong	2	GCS2
		3	GCS3

Due to the complexity of the geodesic and projection-based discretization approaches for directors as well as their high computational cost, discussed in the previous chapter (see Section 3.4), we restrict ourselves to a classical discretization approach. The letter "C" in the abbreviations points to the classical approach, which is frequently used in the literature in combination with the director formulation. It was used with linear shape functions in e.g. [21, 22, 45, 114, 117, 124] and for higher order NURBS elements in [65].

In the case of the Lagrange shape functions, the directors are given at the nodal level in the initialization step. For NURBS they are first computed on the Gauss point and subsequently, the control point values are computed with the algorithm introduced in Section 2.4.2.

To minimize transverse shear locking in combination with NCWp and LCSp elements, for $p = 1, \dots, 3$ selective reduced integration is applied. The Gauss integration is employed for these types of elements. The number of integration points for the internal forces as well as for the integration of the weak constraints for these elements is chosen equal to the element order p. The inertia matrix is fully integrated with $p + 2$ Gauss integration points. In the case of elements with the Gauss-Lobatto-based shape functions, the Lobatto integration is applied. The number of integration points is set to the number of element nodes $p + 1$. Through this approach, the integration is performed at the element nodes. As discussed in Section 3.4.4, this leads, in combination with a strong enforcement of the constraints, to fulfillment of the orthonormality of the directors at the integration points.

The algorithm proposed in [78] was used for mesh refinement in combination with the NURBS-based elements. The order elevation was performed using the definition of k-refinement as introduced in Section 2.4.1. Mesh refinement for elements with Lagrange and Gauss-Lobatto-based functions was performed with equidistant elements over the domain.

The description of the examples in statics in Section 5.6.1 to Section 5.6.4 is taken from Wasmer and Betsch [148]. Adaptations were made to fit into the framework of this work.

5.6.1. Cantilever beam under end moment

The first example deals with the well-known benchmark problem of an initially straight cantilever beam, which is deformed into a circle by an end moment, while being fixed on the other end. A sketch of the problem is shown in Fig. 5.1. The beam has length L and $\bar{\mathbf{m}}$ is an external torque applied at the right end of the beam given by $\bar{\mathbf{m}} = M\mathbf{e}_3$. For $M = 2\pi\frac{EI}{L}$ the beam forms a complete circle, where EI is the bending stiffness of the beam. The beam has an initial length of $L = 1$ and the following stiffness parameters $GA = 1/270$, $EA = 1/100$, $EI = 1/12\; 1 \times 10^{-4}$ and $GI_p = 1/6\; 1 \times 10^{-4}$.

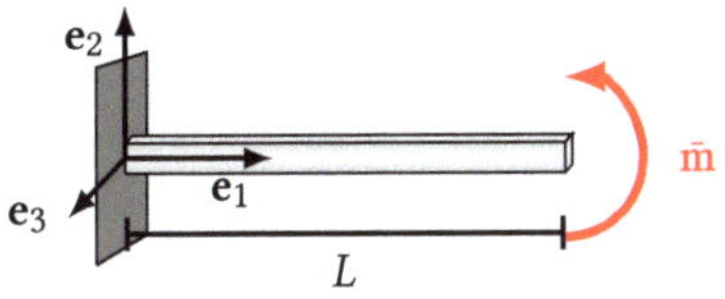

Figure 5.1.: Sketch of the cantilever beam under end moment [148].

The total load is applied in 10 equally spaced incremental steps. As convergence criteria of the Newton solver $||\mathbf{R}|| = 1 \times 10^{-12}$ is used, where $||\mathbf{R}||$ is the 2-norm of the residual vector. The error measure is defined as $e = ||\boldsymbol{\varphi}_{\text{num}}(s = L)||$, where $\boldsymbol{\varphi}_{\text{num}}(s = L)$ is the placement of the tip of the numerical model.

The convergence behavior with respect to h-refinement is shown in Fig. 5.2 for the three types of elements under investigation. All investigated formulations exhibit a very similar behavior for the first two orders. First-order elements display a convergence order of $\mathcal{O}(h^2)$. An increase to second-order shape functions leads to an increase of the convergence order to approximately $\mathcal{O}(h^4)$.

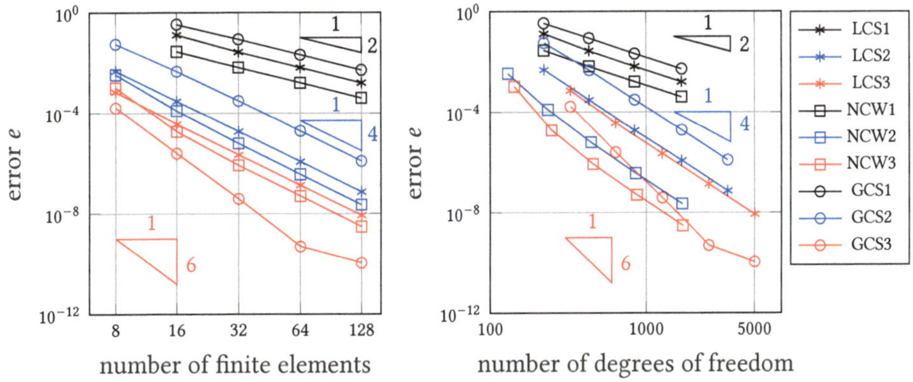

Figure 5.2.: Convergence behavior of the director beam formulation for the roll-up into a circle.

Using third-order elements, however, does not lead to a further increase in the convergence order for the NCW3 and LCS3 elements, as would be expected. NCW3 and LCS3 elements also converge with an order of $\mathcal{O}(h^4)$. This can be explained by the classical finite element discretization. It does not represent the underlying nonlinear manifold $SO(3)$ with sufficient accuracy to achieve optimal convergence behavior. Even for the interpolatory finite elements with Lagrange shape functions and strong enforcement of the orthonormality constraints, the orthonormality is violated at the level of the integration points. This leads to the drop of the convergence order for higher order elements as already discussed in Section 3.4 (cf. [102]).
However, the results obtained with the NURBS shape functions for the first two orders (NCWp, $p = 1, 2$), which are used with weak enforcement of the orthonormality constraints, exhibit a lower error compared to the Lagrange functions and Gauss-Lobatto-based functions for each individual order. Fig. 5.2 reveals for the first-order approach, where both Lagrange and NURBS shape functions coincide, that the weak enforcement of the constraints shows an overall lower error bound. Compared to the other formulations the Gauss-Lobatto-based shape functions, GCW1 and GCW2, lead to the largest error. The larger error can be explained through the limited accuracy of the numerical integration, as a Lobatto integration scheme is chosen in combination with the GCSp elements.

As already mentioned, the GCSp elements show the same order of convergence for order $p = 1, 2$ as the other two approaches, while leading to an overall larger error for each individual element order. Nevertheless, if the element order is increased to $p = 3$ the convergence order of the Gauss-Lobatto-based shape functions increases to approximately $\mathcal{O}(h^6)$.

This increase in the convergence order for the GSC3 elements verifies the correct implementation of the director beam model in the FEM code.

5.6.2. 3D bending and twist

This example is taken from [65]. It consists of an initially straight beam clamped at one end and loaded at the other end with a torque, which is applied in 250 load steps. The torque is applied in such a way that the beam forms a helix with two complete coils. The helix has a radius of $R_0 = 10$ and a height of $h = 50$. The applied torque is given by

$$\bar{\mathbf{m}} = \left(m^1\mathbf{d}_1 + m^3\mathbf{d}_3\right)$$

where $m^1 = -4R_0\frac{\pi^3 r^4}{L^2}$ and $m^3 = \frac{\pi^2 h r^4}{L^2}$. The initial length of the beam is given by $L = \sqrt{1 + c^2}R_0 4\pi$, where $c = \frac{h}{4R_0\pi}$. A slenderness ration ζ is used to define the radius of the beam $r = \frac{L}{2\zeta}$. The beam stiffness parameters are given by $GA = \frac{1}{2}\pi r^2$, $EA = \pi r^2$, $EI = \frac{\pi}{4}r^4$ and $GI_\mathrm{p} = \frac{\pi}{2}r^4$. The straight initial configuration is defined through

$$\boldsymbol{\varphi}(s) = -R_0\mathbf{e}_2 + s\mathbf{d}_3^0$$

where $s \in [0, L]$, and the initial director frame is given by

$$\mathbf{d}_1^0 = R_0\alpha(c\mathbf{e}_1 - \mathbf{e}_3) \qquad \mathbf{d}_2^0 = \mathbf{e}_2 \qquad \mathbf{d}_3^0 = R_0\alpha(\mathbf{e}_1 + c\mathbf{e}_3),$$

where $\alpha = \frac{4\pi}{L}$. The results shown here are computed with a slenderness ratio of $\zeta = 10$. A higher slenderness ratio results in locking effects in the director formulation. In Fig. 5.3 the convergence behavior is presented. The relative error measure is given by

$$e = \frac{||\boldsymbol{\varphi}_{\text{num}}(s = L) - \boldsymbol{\varphi}_{\text{ana}}(s = L)||}{||\boldsymbol{\varphi}_{\text{ana}}(s = L)||},$$

where $\boldsymbol{\varphi}_{\text{ana}}(s = L)$ is the analytic solution for the displacement as given in [65].

The convergence results with respect to mesh refinement are presented in Fig. 5.3. No solution at all could be obtained for first-order Lagrange elements (LCS1) due to strong locking effects. Even choosing a high number of elements does not decrease the locking. This is in accordance with [65]. Furthermore, it is also not possible to find a solution with 16 NCW1 elements due to the divergence of the Newton solver. In contrast to this, a solution for the GCS1 elements can be obtained for all the investigated numbers of elements.
Apart from the issues with the first-order elements, the convergence behavior is in accordance with the previous example. Again we see a convergence order of $\mathcal{O}(h^2)$ for the NCW1 and GCS1 elements and $\mathcal{O}(h^4)$ for all quadratic element types in Fig. 5.3. As before, the GSC3 elements exhibit a convergence order of $\mathcal{O}(h^6)$, while the convergence order of the NCW3 and LCS3 does not increase.
As in the last example, the NURBS-based elements result in a slightly smaller error compared to the Lagrange elements, while for the orders $p = 1, 2$ the Gauss-Lobatto shape functions result in the overall largest error.

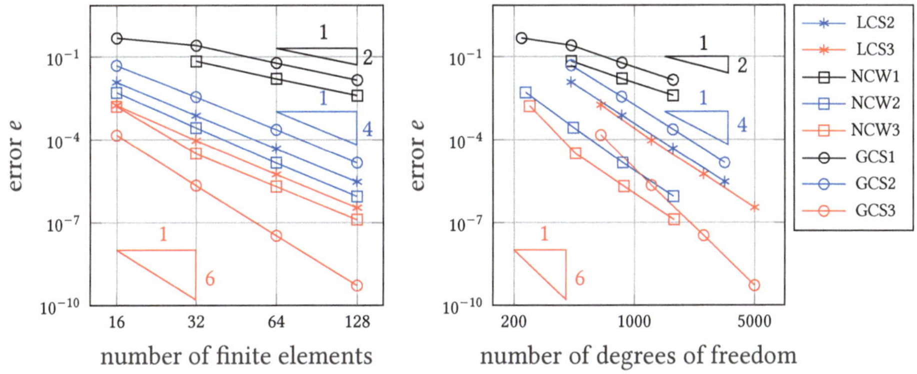

Figure 5.3.: Convergence behavior of the director beam formulation for the roll-up into a helix.

5.6.3. Numerical path-independence test

As mentioned in Chapter 3 and Chapter 4, early discretization approaches of the geometrically exact beam model were neither frame-indifferent nor path-independent. The frame-indifference of the present formulation was shown in Section 5.4.1.1. With this example we verify that the director formulation is also path-independent. As shown in Fig. 5.4, the initial configuration corresponds to a curved beam forming 1/8-th of a circle in the $\mathbf{e}_1$-$\mathbf{e}_3$ plane. The circle has a radius of 100. An external force $\mathbf{F} = F_i \mathbf{e}_i$ is applied

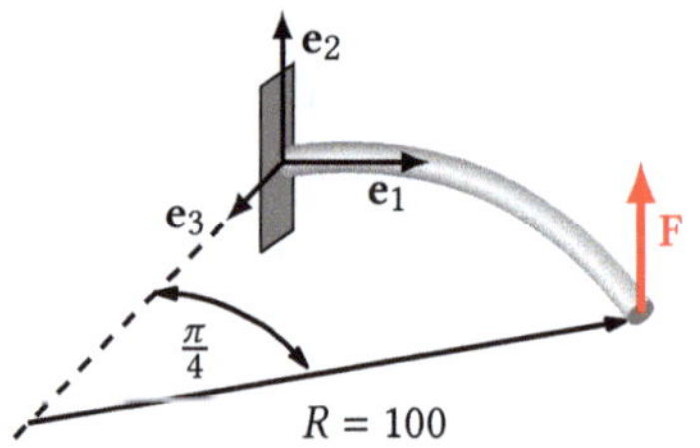

Figure 5.4.: Initial configuration of the curved beam for the path-independence test [148].

in a loading cycle at the tip of the beam. Specific values of the loading cycle are shown in Tab. 5.2. The cycle begins with a force of $\mathbf{F} = \mathbf{0}$. The load level is than varied with $||\mathbf{F}|| = 25$ in each load step. This results in a total of 144 load steps. In the last load step the force is removed so that the initial configuration should be obtained again. The beam's stiffness parameters are given by $GA = 5 \times 10^6$, $EA = 1 \times 10^7$ and $EI = GI_\text{p} = 1/121 \times 10^7$. The convergence criteria ε for the Newton method is set to $||\mathbf{R}|| = 1 \times 10^{-6}$.
The NURBS elements for $p = 2, 3$ are initialized, to represent the 1/8-th circle exactly, for which the algorithm given in Chapter 7.5 in [110] is used. The order elevation or k-refinement is performed with a the algorithm given in Chapter 5.5. in [110], while the h-refinement was performed using the algorithm proposed in [78] as mentioned in Chapter 2. The displacement of the tip in the direction of $\mathbf{e}_2$ is shown in Tab. 5.2 for the various load steps. The presented results are obtained for 32 LCS2, NCW2, and GCW2 elements. Additional results with different number of elements ([8, 16, 32]) and element orders $p = 1, ..., 3$ are presented in the Appendix B. The Gauss-Lobatto shape functions of first-order lead to severe locking. Therefore, the results presented in the Appendix are limited to the computations from the GCS2 and GCS3 elements.

After the whole load cycle is completed the configuration has to coincide with the initial configuration to be path-independent. The results are overall in accordance with the results shown in [21] and the results presented in Chapter 6. It can be observed that the displacement in the direction of $\mathbf{e}_2$ is within a numerical round-off error at the end of the load cycle and, thus, coincides with the initial configuration as expected. The numerical round-off error is by several magnitudes smaller than the convergence criteria ε for the Newton solver. This verifies the path-independence of the director formulation for the investigated element types.

Table 5.2.: Numerical path-independence test: tip displacement in 2-direction corresponding to specific load levels.

load level	tip displacement in 2-direction		
	Lagrange (LCS2)	NURBS (NCW2)	Gauss-Lobatto (GCS2)
$[F_1, \; F_2, \; F_3]$	32 el. $p = 2$	32 el. $p = 2$	32 el. $p = 2$
$[0 \; 0 \; 0]$	0	0	0
$[-600 \; 0 \; 0]$	0	0	0
$[-600 \; 600 \; 0]$	59.8338	59.8755	59.8105
$[-600 \; 600 \; 600]$	38.6974	38.7375	38.6740
$[0 \; 600 \; 600]$	37.5364	37.5810	37.5062
$[0 \; 0 \; 600]$	2.9694×10^{-17}	1.2546×10^{-17}	3.0172×10^{-17}
$[0 \; 0 \; 0]$	$\ll \varepsilon$	$\ll \varepsilon$	$\ll \varepsilon$

Table 5.3.: Residual of Newton solver for last time step discretized with 8 quadratic elements.

Iteration number	LCS2 2-norm of residual	NCW2 2-norm of residual	GCS2 2-norm of residual
1	2.50×10^{1}	2.50×10^{1}	2.50×10^{1}
2	7.05×10^{5}	8.73×10^{5}	4.82×10^{5}
3	2.35×10^{3}	2.84×10^{3}	1.15×10^{3}
4	4.09×10^{2}	5.18×10^{2}	1.02×10^{3}
5	2.25×10^{2}	2.72×10^{2}	7.42×10^{1}
6	4.22×10^{0}	5.34×10^{0}	7.11×10^{0}
7	2.80×10^{-2}	3.42×10^{-2}	3.85×10^{-3}
8	1.75×10^{-7}	1.01×10^{-8}	9.56×10^{-8}

In Tab. 6.4, the convergence behavior of the Newton method for eight quadratic elements is shown for the last loading step. The Newton solver exhibits a very similar convergence behavior for all three element types, which results in an error below the tolerance of the solver in the eighth iteration.

5.6.4. Beam patches with slope discontinuity

In this example, we investigate the capability of the present formulation to deal with discontinuities of the reference curve of the beam. This example has been treated previously in [45, 65, 115]. The initial geometry is shown in Fig. 5.5a. It consists of three beam segments with the length $L = 1$, which are connected rigidly at a $90°$ angle.

The rigid intersection of the beam segments can be either formulated by means of quadratic constraints [124] or, alternatively, without constraints in the framework of the null space matrix [19, 92]. Here, we use the formulation in quadratic constraints.
One end of the beam structure is fixed, while at the other end, a dead load of $\mathbf{F} = -10\mathbf{e}_1 - 10\mathbf{e}_3$ is applied in 30 equally spaced load steps. The beam stiffness parameters are given by $EA = 1 \times 10^4$, $GA = 5 \times 10^3$ and $EI = GI_p = 100/12$. The stopping criteria for Newton's method is set to $||\mathbf{R}|| = 1 \times 10^{-10}$. Each beam segment is discretized with 3 finite elements of order $p = 3$.

The results of the deflection of the tip are plotted over the load level in Fig. 5.5b. All investigated elements are in very good agreement regardless of the employed discretization. No difference can be observed between the three approaches. Moreover, the results are in excellent agreement with the results shown in [45, 65, 115] and the results presented in Chapter 6.

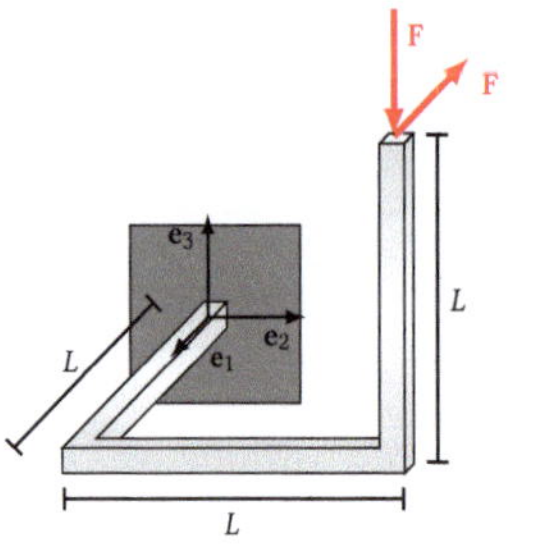

(a) Sketch of the initial configurations of the beam with slope discontinuities [148].

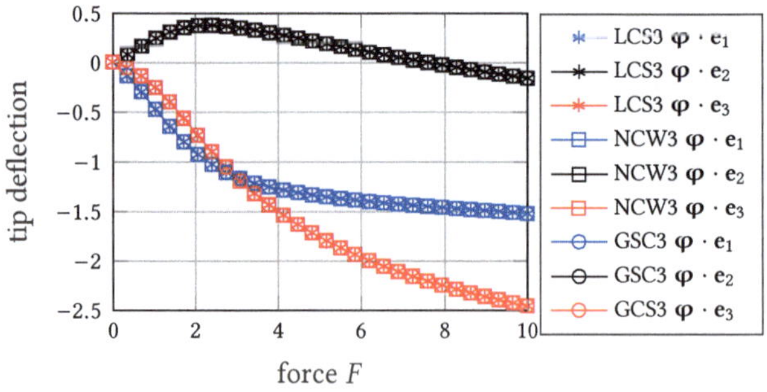

(b) Tip deflection over force for the different element formulations.

Figure 5.5.: Beam patches with slope discontinuity.

5.6.5. Beam with concentrated masses

The first dynamic problem is taken from [22, 124]. In Fig. 5.6 the initial configuration of a beam with concentrated masses is shown. The beam has a total length of $2L$, where $L = 1$. The concentrated masses at the ends are given with $M = 10$ and the mass in the middle with $m = 1$. The beam itself has a mass density per unit length of $A_\rho = 0.27$ and a mass moment of inertia of the cross-section of $M_\rho = 9 \times 10^{-8}$. The beam stiffness parameters are given by $EI = 2.43$, $GI_\mathrm{p} = 586$, $GA = 2.43 \times 10^6$ and $EA = 7.3 \times 10^6$.

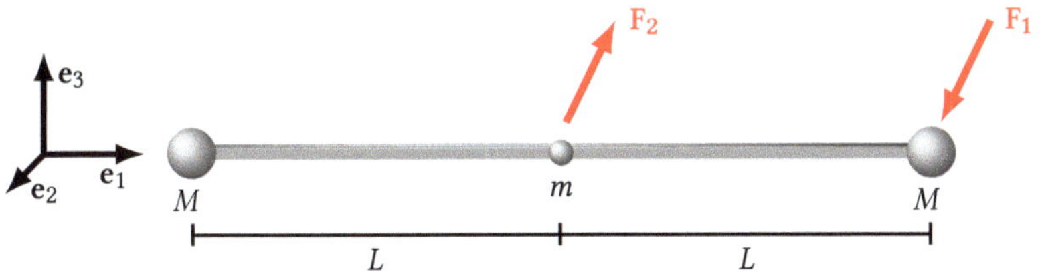

Figure 5.6.: The initial configuration of the beam with concentrated masses.

The system is excited by two time-dependent external dead loads $\mathbf{F}_1$ and $\mathbf{F}_2$

$$\mathbf{F}_\alpha = \mathbf{P}_\alpha f(t), \qquad \text{where} \qquad f(t) = \begin{cases} \frac{1}{2}\left(1 - \cos\left(\frac{2}{3}\pi t\right)\right) & \text{for} \quad t \leq 3 \\ 0 & \text{for} \quad t \geq 3\,. \end{cases}$$

The force vectors $\mathbf{P}_1$ and $\mathbf{P}_2$ are defined to be

$$\mathbf{P}_1 = -1\,\mathbf{e}_1 - 3\,\mathbf{e}_3 \qquad \text{and} \qquad \mathbf{P}_2 = 2\,\mathbf{e}_1 + 4\,\mathbf{e}_3\,.$$

A constant time steps size of $\Delta t = 1 \times 10^{-2}$ is used and the beam is discretized with 22 linear elements for both the Lagrange and NURBS shape functions. As mentioned in the introduction of this chapter, Gauss-Lobatto-based shape functions are not considered.

The simulation results are presented in Fig. 5.7. Results of the LCS1 elements are plotted in gray, while the results obtained with the NURBS shape functions are plotted in colored dashed lines. In Fig. 5.7a the displacement of the mid-span $\varphi^h(s = L)$ is shown. The results of both investigated element types are in very good agreement. No significant differences can be observed.

Fig. 5.7b depicts the total angular momentum of the beam structure over time. Due to the nature of the problem being two-dimensional only the component L_2 can deviate from zero. It can be seen that the angular momentum increases until the load is removed at $t = 3$. In fact, since the energy-momentum time integration scheme, described in Section 5.5, is employed, the angular momentum is constant within the Newton tolerance for $t > 3$. Again, this observation is true regardless of the type of shape function used for the spatial discretization.
The kinetic and potential as well as the total energy of the system over time are shown in Fig. 5.7c. The total energy changes due to the external loading until $t = 3$. For $t > 3$ it is constant up to the Newton tolerance due to the use of the discrete gradient. Even though slight variations between the LCS1 and NCW1 elements can be observed for the kinetic and potential energy, the total energy is in very good agreement for both types of elements. The NCW1 elements result in a slightly smaller total energy. However, the difference (0.4 %) is negligible.

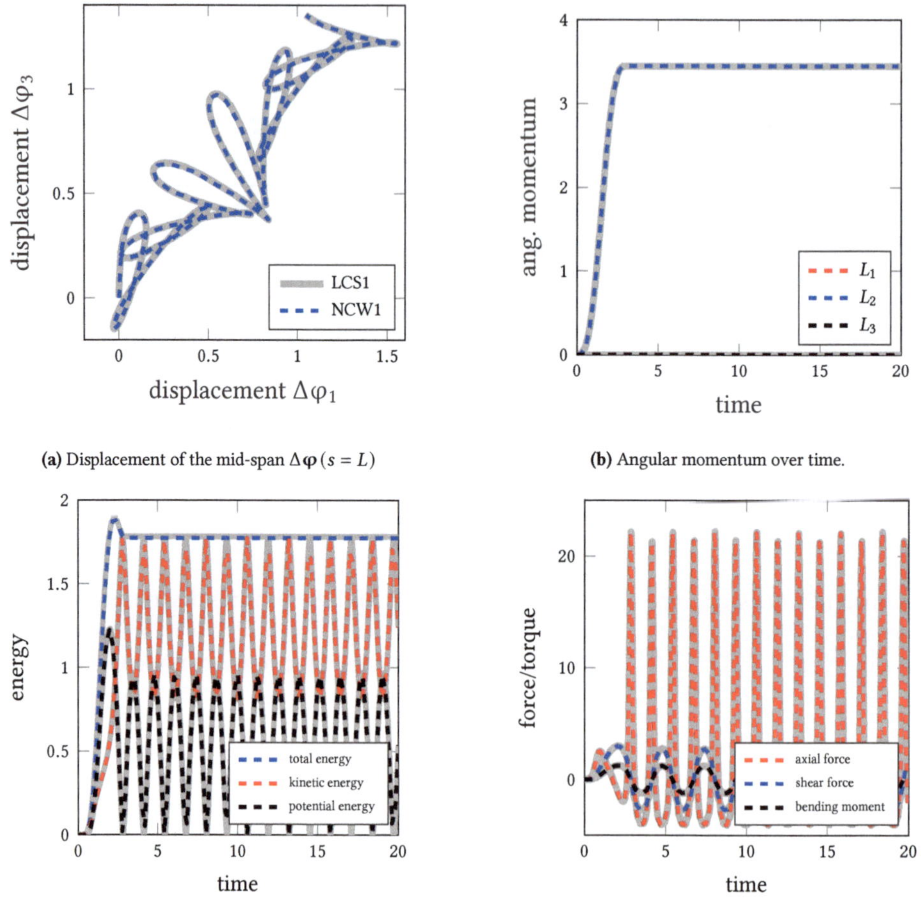

(a) Displacement of the mid-span $\Delta\boldsymbol{\varphi}\,(s = L)$

(b) Angular momentum over time.

(c) Energy over time.

(d) Stress resultants over time.

Figure 5.7.: Results of the example beam with point masses; in grey are the results obtained from the FEM, the dashed, colored lines are the results obtained from the IGA.

The three stress resultants, relevant for the two-dimensional problem, at $s = \frac{L}{2}$ are presented in Fig. 5.7d. Again both types of elements are in line with each other with only minor deviations.
Furthermore, all the presented results are in excellent agreement with the results presented in [22, 124].

5.6.6. Flying Spaghetti

An example often found in the literature is the "Flying Spaghetti", first published by Simo and Vu-Quoc [127]. The initial configuration is a straight beam in the $\mathbf{e}_1$-$\mathbf{e}_2$-plane. The global geometry is defined through the lengths $L_1 = 8$ and $L_2 = 6$. The initial configuration is shown in Fig. 5.8. On one end time-dependent loads, a force $\mathbf{F}(t)$ and a torque $\mathbf{M}(t)$, are applied, which are given by

$$\mathbf{F}(t) = \begin{cases} 8\,\mathbf{e}_1 & \text{for } t \leq 2.5 \\ \mathbf{0} & \text{else} \end{cases} \qquad \text{and} \qquad \mathbf{M}(t) = \begin{cases} -80\,\mathbf{e}_3 & \text{for } t \leq 2.5 \\ \mathbf{0} & \text{else}\,. \end{cases}$$

The structure is chosen to be highly flexible, by setting the beam's stiffness to $GA = 1 \times 10^4$, $EA = 1 \times 10^4$, $EI = 100$ and $GI_\mathrm{p} = 200$. The mass density per unit length is given as $A_\rho = 1$, and the mass moment of inertia as $M_\rho = 10$. A time step of $\Delta t = 0.1$ is used in combination with a convergence criterion of the Newton solver of $||\mathbf{R}|| = 1 \times 10^{-8}$. Again elements with Gauss-Lobatto-based shape functions are not considered.
Snapshots of the Flying Spaghetti's configuration up to $t = 10$ in increments of $\Delta t = 0.5$ are shown in Fig. 5.9, where 10 LCS3 elements are used. The motion is in very good agreement with the frames of motion shown in [127].

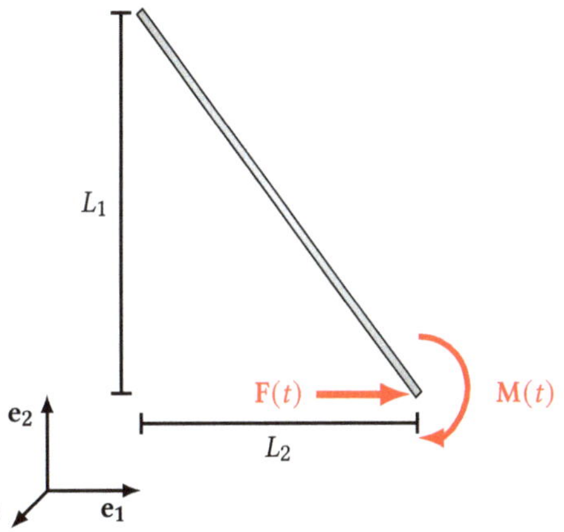

Figure 5.8.: The initial configuration of the flying spaghetti.

The results presented in Fig. 5.10 are obtained for a discretization with 20 LCS3 and NCW3 elements. As in the previous example, the results obtained with the Lagrange elements are shown in gray, while the results of the NURBS elements are plotted with colored dashed lines.

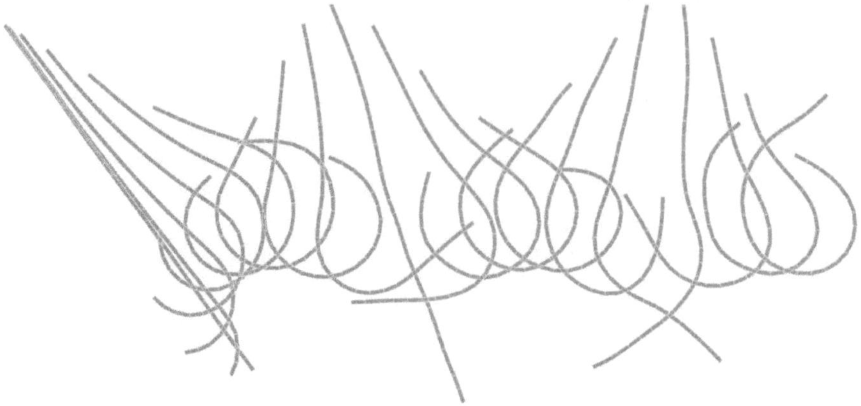

Figure 5.9.: Snapshots of the flight motion of the flying spaghetti (10 LCS3 elements) in increments of $\Delta t = 0.5$ from $t = 0$ to $t = 10$.

The energy history over time is shown in Fig. 5.10a, where the kinetic, potential, and total energies of the beam are plotted. The total energy of the system increases up to $t = 2.5$ when the loads are set equal to zero. For $t > 2.5$ the total energy is constant for both approaches within the Newton tolerance, as expected. The potential and kinetic energy, however, vary over time. Both investigated types of elements are overall in good agreement with only slight variations. As in the previous example, the NCW3 elements result in a slightly smaller total energy. Again the difference (0.75 %) is insignificant.

The total linear momentum of the system is depicted in Fig. 5.10b. The linear momentum P_1 in $\mathbf{e}_1$-direction increases as long as the load is applied. After the load is removed it is conserved. The other two components of the linear momentum, P_2 and P_3, are equal to zero over the whole time domain.
In Fig. 5.10c the total angular momentum of the beam structure is shown. The components L_1 and L_2 of the angular momentum are equal to zero over the total considered time. The angular momentum L_3 decreases until the loads are removed and is constant afterward.

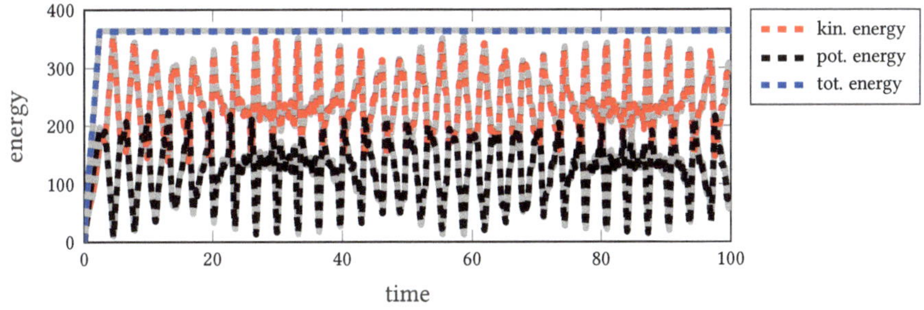

(a) Energy of the Flying Spaghetti over time.

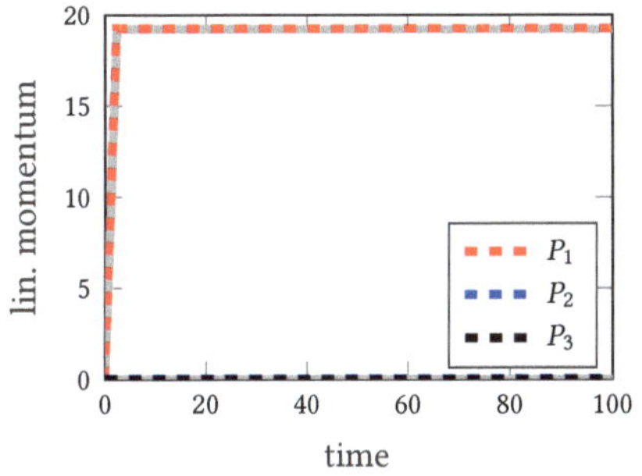

(b) Linear momentum of the Flying Spaghetti over time.

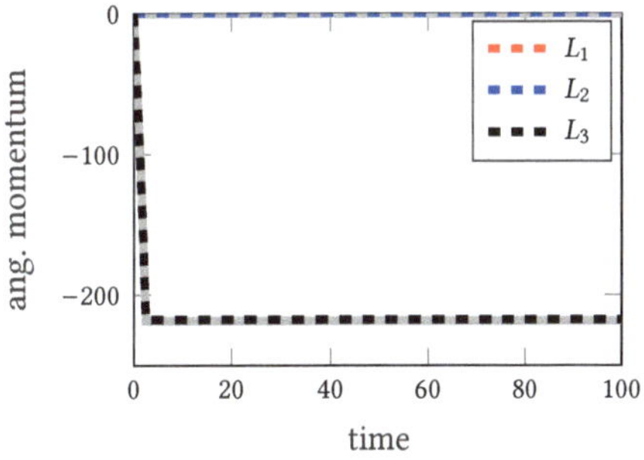

(c) Angular momentum of the Flying Spaghetti over time.

Figure 5.10.: Results for the Flying Spaghetti using 20 cubic elements. The results from the FEM are shown in gray, while the results obtained with NURBS functions are depicted with dashed, colored lines.

5.6.7. Conclusion

The numerical results show that the director formulation does not result in optimal convergence behavior for higher-order elements when a classical discretization approach without additional precautions is taken. Such precautions can be the use of Gauss-Lobatto-based shape functions when the FEM is applied. An equivalent remedy is not apparent for non-interpolatory shape functions such as NURBS. Nevertheless, employing NURBS has shown to be advantageous concerning the accuracy per degree of freedom for lower orders ($p = 1, p = 2$). Furthermore, the director formulation with a classical discretization results in a frame- and path-independent formulation.

Even though there are shortcomings in the convergence rate with a classical discretization, the director formulation can still be employed for dynamic simulations in combination with higher-order elements. However, while the results are still in good agreement with the literature, there might be limited benefit in choosing higher orders ($p \geq 3$) due to the limited convergence rate.

6. The geometrically exact beam in an unit quaternion framework for statics

Abstract: The following chapter covers a formulation of the geometrically exact beam in unit quaternions. At the beginning, a mapping for the unit quaternions into a linear algebra representation is given. In the following, it is used to formulate the geometrically exact beam in quaternions and it is shown a unit quaternion formulation indeed results in frame-indifferent continuous strain measures. Furthermore, the weak formulation of the beam is derived, and the discretization with finite elements is discussed, where a projection-based approach, as introduced in Chapter 3, is employed for the unit quaternion field. It is shown, that such an approach leads to frame-indifference of the discretized strain measures. Furthermore, the discretized contribution of the internal forces, external loading, and the contribution from the unit length constraint is discussed. Two options for the enforcement of the constraints are given. At last, the model is verified by examining the convergence behavior and path-independence using numerical examples.
The following chapter is mainly based on Wasmer and Betsch [148]. Section 6.1 to Section 6.5 are taken from this publication. Adaptations were made to fit into the framework of this work.

As discussed in Chapter 3 rotations can be handled using different concepts, all with advantages and challenges. One very convenient approach is the use of unit quaternions, where details on unit quaternions are given in Section 3.3.3. The most significant advantage of unit quaternions is that they do not result in any singularities when used to parametrize rotations. Furthermore, compared to the minimal set of parameters to describe rotations the use of unit quaternions leads to a moderate increase in the number of unknowns. Thus, quaternions are often used in the context of the geometrically exact beam to represent the orientation of the cross-section area.

Dichmann [40] investigated the geometrically beam in a unit quaternion formulation. A normalization of the quaternions ensures their unit length, to correctly represent the rotational tensor. However, the work does not deal with a finite element formulation.
In the context of beam finite elements, the interpolation of quaternions along with a subsequent normalization is addressed by Romero [114]. However, a detailed treatment

of the resulting finite element formulation is missing. Ghosh and Roy [51] discuss a quaternion-based formulation of the beam in combination with a geodesic element approach for the discretization of the quaternions. An interpolation of the unit quaternions with Lagrangian shape functions and a weak enforcement of the unit constraints is given by Cottanceau *et al.* [30].
Collocation methods based on a quaternion description of the geometrically exact beam can be found in [151, 152]. Weeger *et al.* [149] employ an isogeometric collocation method, whereas Tasora *et al.* [143] use a geodesic elements with unit quaternion in the IGA as a Galerkin method. However, again a detailed discussion of the discretization approach is not given.

In this chapter, we introduce the beams equation in a unit quaternion description. In detail, the geometrically exact beam formulation in a unit quaternion formulation is presented. In contrast to most mentioned publications, which use Lagrange elements, we employ the isogeometric concept with NURBS in this chapter. A detailed application of the projection-based approach for Lagrange and NURBS shape functions is given. We focus in this chapter on a purely quasi-static approach.
To eliminate the redundancy introduced by the use of unit quaternions, two different approaches are investigated. Both methods rely on Lagrange multipliers, however, one of the methods is well suited to be used in a size reduction approach allowing the equations to be solved on the minimal set of coordinates.
A formulation of the geometrically exact beam model in unit quaternions is, of course, no new approach. However, to the best knowledge of the author, a projection-based approach for unit quaternions was never investigated in combination with the Isogeometric Analysis. Furthermore, no in-depth discussion of such an approach for Lagrange elements with a presentation of the convergence behavior is given in the literature. The goal of this chapter is to fill this gap.

This chapter is structured as follows. At the beginning, a linear algebra representation of the quaternion product is introduced in Section 6.1. In Section 6.2 we present the kinematics of the geometrically exact beam in terms of the unit quaternions. The subsequent Section 6.4 deals with the application of the Finite Element Method to the resulting weak form. A projection-based approach is employed, where we show the frame-indifference of the proposed discretization. In Section 6.5 we show the results for both approaches alongside examples from Chapter 5. The convergence behavior of the proposed discretization approaches as well as their path-independence are shown. Besides the projection-based approach, we discuss the results of the classical discretization approach.

6.1. Quaternions in a linear algebra representation

For what follows it is useful to rewrite the quaternion product as a tensor multiplication. In this work we use the notation as introduced in [20]. The quaternion product of two quaternions

$$\mathbb{v} = \mathbb{q} \circ \mathbb{p} \tag{6.1}$$

can be rewritten as

$$\mathbb{v} = \mathbf{Q}_l(\mathbb{q}) \cdot \mathbb{p} = \mathbf{Q}_r(\mathbb{p}) \cdot \mathbb{q} \,. \tag{6.2}$$

The mappings $\mathbf{Q}_l(\mathbb{q})$ and $\mathbf{Q}_r(\mathbb{q})$ from $\mathbb{H} \mapsto \mathbb{R}^{4\times4}$ are given by

$$\mathbf{Q}_l(\mathbb{q}) = \begin{bmatrix} \mathbb{q} & \mathbf{G}(\mathbb{q})^\top \end{bmatrix}, \qquad \mathbf{Q}_r(\mathbb{q}) = \begin{bmatrix} \mathbb{q} & \mathbf{E}(\mathbb{q})^\top \end{bmatrix}. \tag{6.3}$$

Above we introduced two mappings $\mathbf{E}(\mathbb{q}), \mathbf{G}(\mathbb{q}) : \mathbb{H} \mapsto \mathbb{R}^{3\times4}$

$$\mathbf{G}(\mathbb{q}) = \begin{bmatrix} -\mathbf{q} & q_0\mathbf{I}_3 - \hat{\mathbf{q}} \end{bmatrix}, \qquad \mathbf{E}(\mathbb{q}) = \begin{bmatrix} -\mathbf{q} & q_0\mathbf{I}_3 + \hat{\mathbf{q}} \end{bmatrix}. \tag{6.4}$$

A rotational tensor Eq. (3.46) can now be written in the alternative form

$$\mathbf{R}(\mathbb{q}) = \mathbf{E}(\mathbb{q}) \cdot \mathbf{G}(\mathbb{q})^\top \tag{6.5}$$

for any $\mathbb{q} \in S^3$. Further useful algebraic relationships involving the matrices $\mathbf{E}(\mathbb{p})$ and $\mathbf{G}(\mathbb{p})$ for any $\mathbb{p} \in \mathbb{H}$ can be summarized as follows:

$$\begin{aligned} \mathbf{E}(\mathbb{p}) \cdot \mathbb{p} &= \mathbf{G}(\mathbb{p}) \cdot \mathbb{p} = \mathbf{0}, \\ \mathbf{E}(\mathbb{p}) \cdot \mathbf{E}(\mathbb{p})^\top &= \mathbf{G}(\mathbb{p}) \cdot \mathbf{G}(\mathbb{p})^\top = ||\mathbb{p}||^2\, \mathbf{I}_3, \\ \mathbf{E}(\mathbb{p})^\top \cdot \mathbf{E}(\mathbb{p}) &= \mathbf{G}(\mathbb{p})^\top \cdot \mathbf{G}(\mathbb{p}) = ||\mathbb{p}||^2\, \mathbf{I}_4 - \mathbb{p} \otimes \mathbb{p} \,. \end{aligned} \tag{6.6}$$

Moreover, for $\mathbb{p}, \mathbb{u} \in \mathbb{H}$

$$\begin{aligned} \mathbf{E}(\mathbb{p}) \cdot \mathbb{u} &= -\mathbf{E}(\mathbb{u}) \cdot \mathbb{p}, \\ \mathbf{G}(\mathbb{p}) \cdot \mathbb{u} &= -\mathbf{G}(\mathbb{u}) \cdot \mathbb{p}, \\ \mathbf{E}(\mathbb{p}) \cdot \mathbf{G}(\mathbb{u})^\top &= \mathbf{E}(\mathbb{u}) \cdot \mathbf{G}(\mathbb{p})^\top . \end{aligned} \tag{6.7}$$

We further define the skew-symmetric matrix $\overset{+}{\mathbf{v}} \in \mathbb{R}^{4\times4}$ mapping any vector $\mathbf{v} : \mathbb{R}^3 \longrightarrow \mathbb{R}^{4\times4}$ such that the relationship

$$\mathbf{E}(\mathbb{p})^\top \cdot \mathbf{v} = \overset{+}{\mathbf{v}} \cdot \mathbb{p} \quad \text{where} \quad \overset{+}{\mathbf{v}} = \begin{bmatrix} 0 & -\mathbf{v}^\top \\ \mathbf{v} & \hat{\mathbf{v}} \end{bmatrix} \tag{6.8}$$

holds for any $\mathbb{p} \in \mathbb{H}$ and $\mathbf{v} \in \mathbb{R}^3$.

6.2. Beam kinematics of the unit quaternion formulation

As shown in Chapter 4 the three directors are assumed to be mutually orthonormal so that the director frame can be expressed in terms of a rotation tensor $\mathbf{R}(s) \in SO(3)$ via

$$\mathbf{d}_i(s) = \mathbf{R}(s) \cdot \mathbf{e}_i \tag{6.9}$$

for $i = 1, \dots, 3$, where the Cartesian base vectors $\mathbf{e}_i \in \mathbb{R}^3$ form an inertial reference frame. The configuration space of the geometrically exact beam model can be written as

$$\tilde{\mathbb{Q}} = \{(\boldsymbol{\varphi}, \mathbf{R}) : [0, L] \to \mathbb{R}^3 \times SO(3)\}\,. \tag{6.10}$$

In the chapter, we employ unit quaternions to represent the rotation tensor. Accordingly, instead of Eq. (6.9), the directors are expressed in terms of the unit quaternion $\mathrm{q}(s) \in S^3$ through the relationship

$$(0, \mathbf{d}_i(s)) = \mathrm{q}(s) \circ (0, \mathbf{e}_i) \circ \bar{\mathrm{q}}(s)\,. \tag{6.11}$$

Alternatively, we may write $\mathbf{d}_i = \mathbf{R}(\mathrm{q}) \cdot \mathbf{e}_i$, where the rotation tensor $\mathbf{R} : S^3 \mapsto SO(3)$ assumes the form Eq. (3.46). Correspondingly, the configuration space of the beam is now given by

$$\mathbb{Q}_{\mathrm{q}} = \{(\boldsymbol{\varphi}, \mathrm{q}) : [0, L] \to \mathbb{R}^3 \times S^3\}\,. \tag{6.12}$$

Since $\mathrm{q} \in S^3$ implies the unit-length condition $\mathrm{q} \cdot \mathrm{q} = 1$, the derivative of this condition with respect to the arc-length yields $\mathrm{q}_{,s} \cdot \mathrm{q} = 0$, so that $\mathrm{q}_{,s} \in T_{\mathrm{q}}S^3$, the tangent space of S^3 at $\mathrm{q} \in S^3$ given by

$$T_{\mathrm{q}}S^3 = \{\mathrm{v} \in \mathbb{H} \mid \mathrm{q} \cdot \mathrm{v} = 0\}\,. \tag{6.13}$$

6.2.1. Strain measures using quaternions

In this section we provide the strain measures of the geometrically exact beam model based on the parametrization $(\boldsymbol{\varphi}, \mathrm{q}) \in \mathbb{Q}_{\mathrm{q}}$. We start with the strains associated with bending and twist which can be arranged in vector $\mathbf{K} \in \mathbb{R}^3$. Vector $\mathbf{K}$ is defined as axial vector associated with the skew-symmetric tensor[1]

$$\widehat{\mathbf{K}} = \mathbf{R}^\top \cdot \mathbf{R}_{,s}\,. \tag{6.14}$$

Since

$$\mathbf{R}_{,s} = \left(\mathbf{E}(\mathrm{q}) \cdot \mathbf{G}(\mathrm{q})^\top\right)_{,s} = 2\mathbf{E}(\mathrm{q}) \cdot \mathbf{G}(\mathrm{q}_{,s})^\top \tag{6.15}$$

[1] For simplicity, we content ourselves here with initially straight beams. The extension to initially curved beams is straightforward, see, for example, Chapter 4 or [21, 65].

Eq. (6.14) yields

$$\widehat{\mathbf{K}} = 2\mathbf{G}(\mathbb{q}) \cdot \mathbf{G}(\mathbb{q}_{,s})^{\top}, \tag{6.16}$$

where Eq. (6.7)$_3$ and Eq. (6.6)$_1$ have been used. It can be shown that the axial vector corresponding to Eq. (6.16) is given by

$$\boxed{\mathbf{K} = 2\mathbf{G}(\mathbb{q}) \cdot \mathbb{q}_{,s}}\,. \tag{6.17}$$

For completeness this is shown in Appendix C. Note that strain measure Eq. (6.17) can also be written as

$$(0, \mathbf{K}) = 2\mathbf{Q}_l(\mathbb{q})^{\top} \cdot \mathbb{q}_{,s} = 2\bar{\mathbb{q}} \circ \mathbb{q}_{,s}\,. \tag{6.18}$$

The second strain measure of the geometrically exact beam model is defined by $\mathbf{\Gamma} = \mathbf{R}^{\top} \cdot \boldsymbol{\varphi}_{,s} - \mathbf{e}_3$ and accounts for transverse shear and normal strain. The representation of the rotation group by means of unit quaternions leads to

$$\boxed{\mathbf{\Gamma} = \mathbf{R}(\mathbb{q})^{\top} \cdot \boldsymbol{\varphi}_{,s} - \mathbf{e}_3}\,, \tag{6.19}$$

where the rotation tensor has been introduced in Eq.(6.5). Alternatively, the strain measure $\mathbf{\Gamma}$ in Eq. (6.19) can be written as

$$(0, \mathbf{\Gamma}) = \bar{\mathbb{q}} \circ (0, \boldsymbol{\varphi}_{,s}) \circ \mathbb{q} - (0, \mathbf{e}_3)\,. \tag{6.20}$$

6.2.1.1. Frame-indifference of the strain measures

The strain measures of the beam theory at hand are invariant under rigid motions. The so-called frame-indifference (or objectivity) of the strains can be shown by considering a superposed rigid motion of the beam defined by

$$\begin{aligned} (0, \boldsymbol{\varphi}^{\#}) &= \mathbb{r} \circ (0, \boldsymbol{\varphi}) \circ \bar{\mathbb{r}} + (0, \mathbf{c})\,, \\ \mathbb{q}^{\#} &= \mathbb{r} \circ \mathbb{q}\,. \end{aligned} \tag{6.21}$$

Here, $\mathbb{r} \in S^3$ represents an arbitrary rotation while $\mathbf{c} \in \mathbb{R}^3$ represents an arbitrary translation. Inserting Eq. (6.21)$_1$ into Eq. (6.18) yields

$$\begin{aligned} (0, \mathbf{K}^{\#}) &= 2\bar{\mathbb{q}}^{\#} \circ \mathbb{q}^{\#}_{,s} \\ &= 2\bar{\mathbb{q}} \circ \bar{\mathbb{r}} \circ \mathbb{r} \circ \mathbb{q}_{,s} \\ &= 2\bar{\mathbb{q}} \circ \mathbb{q}_{,s} \\ &= (0, \mathbf{K})\,, \end{aligned} \tag{6.22}$$

which shows the invariance of $\mathbf{K}$ under rigid motions. Similarly, substituting from Eq.(6.21) into Eq. (6.20) yields

$$\begin{aligned} (0, \mathbf{\Gamma}^{\#}) &= \bar{\mathbb{q}}^{\#} \circ (0, \boldsymbol{\varphi}^{\#}_{,s}) \circ \mathbb{q}^{\#} - (0, \mathbf{e}_3) \\ &= \bar{\mathbb{q}} \circ \bar{\mathbb{r}} \circ \mathbb{r} \circ (0, \boldsymbol{\varphi}_{,s}) \circ \bar{\mathbb{r}} \circ \mathbb{r} \circ \mathbb{q} - (0, \mathbf{e}_3) \\ &= \bar{\mathbb{q}} \circ (0, \boldsymbol{\varphi}_{,s}) \circ \mathbb{q} - (0, \mathbf{e}_3) \\ &= (0, \mathbf{\Gamma})\,. \end{aligned} \tag{6.23}$$

In the present work we aim at a beam finite element formulation that inherits the frame-indifference of the strain measures from the underlying continuous formulation, cf. Chapter 4 and [21, 33].

6.3. Weak form of the unit quaternion formulation

The weak form of the equilibrium problem pertaining to the present beam formulation can be written in the standard form

$$G_{\text{int}}(\boldsymbol{\varphi}, \mathbb{q}; \delta\boldsymbol{\varphi}, \delta\mathbb{q}) = G_{\text{ext}}(\boldsymbol{\varphi}, \mathbb{q}; \delta\boldsymbol{\varphi}, \delta\mathbb{q}) \tag{6.24}$$

for $(\boldsymbol{\varphi}, \mathbb{q}) \in \mathbb{Q}$ and arbitrary admissible variations $(\delta\boldsymbol{\varphi}, \delta\mathbb{q}) \in T_{(\boldsymbol{\varphi},\mathbb{q})}\mathbb{Q}$ lying in the tangent space

$$T_{(\boldsymbol{\varphi},\mathbb{q})}\mathbb{Q} = \{(\delta\boldsymbol{\varphi}, \delta\mathbb{q}) : [0, L] \to \mathbb{R}^3 \times T_{\mathbb{q}}S^3\}\,. \tag{6.25}$$

The virtual work contribution of the internal forces results from

$$G_{\text{int}}(\boldsymbol{\varphi}, \mathbb{q}; \delta\boldsymbol{\varphi}, \delta\mathbb{q}) = \int_0^L \delta\boldsymbol{\Gamma} \cdot \mathbf{N} + \delta\mathbf{K} \cdot \mathbf{M}\, \mathrm{d}s\,. \tag{6.26}$$

For the stress resultants $\mathbf{N}$ and $\mathbf{M}$ Saint-Vernant type constitutive laws of the form $\mathbf{N} = \mathbf{D}_1 \cdot \boldsymbol{\Gamma}$ and $\mathbf{M} = \mathbf{D}_2 \cdot \mathbf{K}$ are assumed. The stiffness tensors are given by $\mathbf{D}_1 = \text{diag}(GA_1, GA_2, EA)$ and $\mathbf{D}_2 = \text{diag}(EI_1, EI_2, GI_\text{p})$, respectively. Starting from the strain measures Eq. (6.17) and Eq. (6.19), a straightforward calculation taking into account Eq. (6.7) yields

$$\begin{aligned} G_{\text{int}}(\boldsymbol{\varphi}, \mathbb{q}; \delta\boldsymbol{\varphi}, \delta\mathbb{q}) = \int_0^L & \left(\delta\boldsymbol{\varphi}_{,s} \cdot \mathbf{E}(\mathbb{q}) + 2\boldsymbol{\varphi}_{,s} \cdot \mathbf{E}(\delta\mathbb{q})\right) \cdot \mathbf{G}(\mathbb{q})^\top \cdot \mathbf{N} \\ & + 2\left(\delta\mathbb{q}_{,s} \cdot \mathbf{G}(\mathbb{q})^\top - \delta\mathbb{q} \cdot \mathbf{G}(\mathbb{q}_{,s})^\top\right) \cdot \mathbf{M}\, \mathrm{d}s\,. \end{aligned} \tag{6.27}$$

The virtual work contribution of the external loading is given by

$$G_{\text{ext}}(\boldsymbol{\varphi}, \mathbb{q}; \delta\boldsymbol{\varphi}, \delta\mathbb{q}) = \int_0^L \delta\boldsymbol{\varphi} \cdot \bar{\mathbf{n}} + \delta\mathbb{q} \cdot 2\mathbf{E}(\mathbb{q})^\top \cdot \bar{\mathbf{m}}\, \mathrm{d}s\,, \tag{6.28}$$

where $\bar{\mathbf{n}} \in \mathbb{R}^3$ and $\bar{\mathbf{m}} \in \mathbb{R}^3$ are prescribed external forces and torques, respectively, acting along the centerline of the beam. Note that, for simplicity, boundary terms have been neglected in the above description.

6.4. Finite element formulation of the unit quaternion formulation

We aim at a finite element formulation of the present beam model that both inherits the objectivity of the strain measures and preserves the unit-length of the quaternion field governing the rotation of the director frame throughout the discrete beam formulation. For that purpose the unit quaternion field is approximated by

$$\mathbb{q}^h(s) = \frac{\mathbb{p}^h(s)}{\|\mathbb{p}^h(s)\|}, \quad \text{where} \quad \mathbb{p}^h(s) = \sum_{i=1}^{n} N_i(s)\mathbb{p}_i, \tag{6.29}$$

which is the projection-based approach introduced in Section 3.4.3. As basis functions $N_i(s)$ we employ either NURBS shape functions (see Section 2.3 for further details) or standard Lagrange shape functions. Correspondingly, $\mathbb{p}_i \in \mathbb{H}$ are either the values at the control points or the nodal values of the discretized quaternions, respectively.

Note that the normalization of the discretized quaternion field Eq. (6.29) ensures that the discrete quaternions belong pointwise to the set of unit quaternions which are employed to parametrize the rotation manifold. That is, $\mathbb{q}^h \in S^3$ holds throughout the discrete beam formulation.

Taking the derivative of Eq. (6.29) with respect to the arc-length yields

$$\mathbb{q}^h_{,s} = \frac{\mathbb{p}^h_{,s}}{\|\mathbb{p}^h\|} - \frac{1}{\|\mathbb{p}^h\|^3}\mathbb{p}^h \cdot \left(\mathbb{p}^h \cdot \mathbb{p}^h_{,s}\right) = \frac{1}{\|\mathbb{p}^h\|}\mathbf{P}_{TS^3}(\mathbb{q}^h) \cdot \mathbb{p}^h_{,s}, \tag{6.30}$$

where relation Eq. (3.33) has been taken into account. Moreover, $\mathbf{P}_{TS^3}(\mathbb{q}^h)$ is a projector onto the tangent space of the unit sphere given by [2]

$$\mathbf{P}_{TS^3}(\mathbb{q}^h) = \mathbf{I}_4 - \mathbb{q}^h \otimes \mathbb{q}^h \tag{6.31}$$

as discussed in Section 3.4.3. Thus Eq. (6.30) ensures that $\mathbb{q}^h_{,s} \in T_{\mathbb{q}^h}S^3$. Similarly, variations of $\mathbb{q}^h \in S^3$ can be written in the form

$$\delta\mathbb{q}^h = \frac{1}{\|\mathbb{p}^h\|}\mathbf{P}_{TS^3}(\mathbb{q}^h) \cdot \delta\mathbb{p}^h \tag{6.32}$$

ensuring that $\delta\mathbb{q}^h \in T_{\mathbb{q}^h}S^3$ for any

$$\delta\mathbb{p}^h(s) = \sum_{i=1}^{n} N_i(s)\delta\mathbb{p}_i. \tag{6.33}$$

Concerning the finite element approximation of the center line of the beam we apply the standard discretization

$$\boldsymbol{\varphi}^h(s) = \sum_{i=1}^{n} N_i(s)\boldsymbol{\varphi}_i \tag{6.34}$$

along with

$$\delta\boldsymbol{\varphi}^h(s) = \sum_{i=1}^{n} N_i(s)\delta\boldsymbol{\varphi}_i \tag{6.35}$$

for the corresponding test function. Again, $\boldsymbol{\varphi}_i, \delta\boldsymbol{\varphi}_i \in \mathbb{R}^3$ play the role of either the respective values at the control points, or the nodal values. In the sequel, we shall refer to the i-th point of the discretization as control/nodal point.

We emphasize that the present discretization approach respects the configuration space of the geometrically exact beam model in the sense that $(\boldsymbol{\varphi}^h, \mathbb{q}^h) \in \mathbb{Q}_{\mathbb{q}}$ and $(\delta\boldsymbol{\varphi}^h, \delta\mathbb{q}^h) \in T_{(\boldsymbol{\varphi}^h, \mathbb{q}^h)}\mathbb{Q}_{\mathbb{q}}$, respectively.

6.4.1. Discrete strain measures

The discretized strain measures result from inserting the approximations Eq. (6.29) and Eq. (6.34) into the strain measures Eq. (6.17) and Eq. (6.19). Accordingly, strain measure Eq. (6.17) leads to the discrete counterpart

$$\mathbf{K}^h = 2\mathbf{G}(\mathbb{q}^h) \cdot \mathbb{q}^h_{,s} \,. \tag{6.36}$$

Inserting expression Eq. (6.30) for $\mathbb{q}^h_{,s}$ into Eq. (6.36) and taking into account Eq. $(6.6)_1$ eventually leads to the discretized strain measure

$$\mathbf{K}^h = \frac{2}{\|\mathbb{p}^h\|^2}\mathbf{G}(\mathbb{p}^h) \cdot \mathbb{p}^h_{,s} \,. \tag{6.37}$$

Similarly, the discretized version of strain measure Eq. (6.19) can be written in the form

$$\boldsymbol{\Gamma}^h = \frac{1}{\|\mathbb{p}^h\|^2}\mathbf{R}(\mathbb{p}^h)^\top \cdot \boldsymbol{\varphi}^h_{,s} - \mathbf{e}_3 \,. \tag{6.38}$$

The discrete strain measures $\mathbf{K}^h$ and $\boldsymbol{\Gamma}^h$ inherit the frame-indifference of the underlying continuous strain measures, as will be shown below.

6.4.1.1. Frame-indifference of the discrete strain measures

Similar to the continuous case, frame-indifference of the discrete strain measures can be shown by considering a superposed rigid motion of the discrete model of the beam. Despite the non-interpolatory nature of the NURBS basis, affine transformations of NURBS curves can be achieved by applying these transformations directly to the control points [32], which is in complete analogy to the use of standard Lagrange shape functions. Correspondingly, a rigid motion of the control/nodal points is characterized by

$$\begin{aligned} \boldsymbol{\varphi}_i^{\#} &= \mathbf{R}(\mathbb{r}) \cdot \boldsymbol{\varphi}_i + \mathbf{c} \\ \mathbb{p}_i^{\#} &= \mathbb{r} \circ \mathbb{p}_i = \mathbf{Q}_l(\mathbb{r}) \cdot \mathbb{p}_i \,. \end{aligned} \tag{6.39}$$

As before, $\mathbb{r} \in S^3$ represents an arbitrary rotation with associated rotation tensor $\mathbf{R}(\mathbb{r}) = \mathbf{E}(\mathbb{r}) \cdot \mathbf{G}(\mathbb{r})^\top$. It can be easily verified that Eq. $(6.39)_2$ implies

$$\mathbb{p}^{h^\#} = \mathbb{r} \circ \mathbb{p}^h = \mathbf{Q}_l(\mathbb{r}) \cdot \mathbb{p}^h \tag{6.40}$$

and

$$\left\|\mathbb{p}^{h^\#}\right\|^2 = \mathbb{p}^{h^\#} \cdot \mathbb{p}^{h^\#} = \mathbb{p}^h \cdot \mathbf{Q}_l(\mathbb{r})^\top \cdot \mathbf{Q}_l(\mathbb{r}) \cdot \mathbb{p}^h = \mathbb{p}^h \cdot \mathbb{p}^h = \left\|\mathbb{p}^h\right\|^2 , \tag{6.41}$$

where $\mathbf{Q}_l(\mathbb{r})^\top \cdot \mathbf{Q}_l(\mathbb{r}) = \mathbf{I}_4$ has been used. Now Eq. (6.37) gives rise to

$$\begin{aligned} \mathbf{K}^{h^\#} &= \frac{2}{\|\mathbb{p}^{h^\#}\|^2} \mathbf{G}(\mathbb{p}^{h^\#}) \cdot \mathbb{p}^{h^\#}_{,s} \\ &= \frac{2}{\|\mathbb{p}^h\|^2} \mathbf{G}(\mathbb{r} \circ \mathbb{p}^h) \cdot \mathbf{Q}_l(\mathbb{r}) \cdot \mathbb{p}^h_{,s} . \end{aligned} \tag{6.42}$$

It can be verified by a straightforward calculation that the identity

$$\mathbf{G}(\mathbb{r} \circ \mathbb{p}^h) = \mathbf{G}(\mathbb{p}^h) \cdot \mathbf{Q}_l(\mathbb{r})^\top \tag{6.43}$$

holds. Using the last equation along with the orthogonality of $\mathbf{Q}_l(\mathbb{r})$ leads to the result $\mathbf{K}^{h^\#} = \mathbf{K}^h$, which confirms the invariance of the discrete strain measure $\mathbf{K}^h$ under rigid motions. Similarly, we consider

$$\Gamma^{h^\#} = \frac{1}{\|\mathbb{p}^{h^\#}\|^2} \mathbf{R}(\mathbb{p}^{h^\#})^\top \cdot \boldsymbol{\varphi}^{h^\#}_{,s} - \mathbf{e}_3 , \tag{6.44}$$

where Eq. (6.34) together with Eq. $(6.39)_1$ implies

$$\boldsymbol{\varphi}^{h^\#}_{,s} = \mathbf{R}(\mathbb{r}) \cdot \sum_{i=1}^{n} N_{i,s} \boldsymbol{\varphi}_i + \mathbf{c} \sum_{i=1}^{n} N_{i,s} = \mathbf{R}(\mathbb{r}) \cdot \boldsymbol{\varphi}^h_{,s} . \tag{6.45}$$

Here, the partition of unity property $\sum_{i=1}^n N_i = 1$ has been used which holds for both the NURBS and the Lagrange shape functions. We thus obtain

$$\Gamma^{h^\#} = \frac{1}{\|\mathbb{p}^h\|^2} \cdot \mathbf{R}(\mathbb{r} \circ \mathbb{p}^h)^\top \cdot \mathbf{R}(\mathbb{r}) \cdot \boldsymbol{\varphi}^h_{,s} . \tag{6.46}$$

Since the identity

$$\mathbf{R}(\mathbb{r} \circ \mathbb{p}) = \mathbf{R}(\mathbb{r}) \cdot \mathbf{R}(\mathbb{p}) \tag{6.47}$$

holds for any $\mathbb{p} \in \mathbb{H}$, we eventually obtain

$$\begin{aligned} \Gamma^{h^\#} &= \frac{1}{\|\mathbb{p}^h\|^2} \mathbf{R}(\mathbb{p}^h)^\top \cdot \mathbf{R}(\mathbb{r})^\top \cdot \mathbf{R}(\mathbb{r}) \cdot \boldsymbol{\varphi}^h_{,s} \\ &= \frac{1}{\|\mathbb{p}^h\|^2} \mathbf{R}(\mathbb{p}^h)^\top \cdot \boldsymbol{\varphi}^h_{,s} \\ &= \Gamma^h , \end{aligned} \tag{6.48}$$

which corroborates the frame-indifference of the discrete strain measure Γ^h.

6.4.2. Discrete weak form

The finite element formulation essentially follows from inserting the finite element approximations described above into weak form Eq. (6.24). Although we ensure that $(\mathbb{q}^h, \delta\mathbb{q}^h) \in TS^3$ for all $s \in [0, L]$ from the outset, we still have to take care of the redundancy of $(\mathbb{p}_i, \delta\mathbb{p}_i) \in \mathbb{H} \times \mathbb{H}$. In essence, this redundancy occurs due to the fact that the quaternion variables in the control/nodal points belong to a 4-dimensional space, whereas $\dim(S^3) = 3$. To account for the redundancy, we impose the unit-length constraint on the quaternion field $\mathbb{p}^h \in \mathbb{H}$ by applying the method of Lagrange multipliers. Accordingly, we introduce the discretized weak form

$$G_{\text{int}}(\boldsymbol{\varphi}^h, \mathbb{q}^h; \delta\boldsymbol{\varphi}^h, \delta\mathbb{q}^h) + G_\lambda(\mathbb{p}^h, \lambda^h; \delta\mathbb{p}^h) = G_{\text{ext}}(\boldsymbol{\varphi}^h, \mathbb{q}^h; \delta\boldsymbol{\varphi}^h, \delta\mathbb{q}^h), \tag{6.49}$$

where G_{int} and G_{ext} are given by Eq. (6.27) and Eq. (6.28), respectively. Furthermore, G_λ accounts for the unit-length constraint on $\mathbb{p}^h \in \mathbb{H}$. In this connection, λ^h represent the contribution of the Lagrange multipliers. Two alternative versions of G_λ will be introduced in Section 6.4.2.2.

The contribution of the internal forces to the weak form, G_{int} in Eq. (6.49), can be obtained by starting from the continuous expression for G_{int}, Eq. (6.27), and making use of Eq. (6.29), Eq. (6.30) and Eq. (6.32). Accordingly, a straightforward calculation yields

$$\begin{aligned} G^h_{\text{int}}(\boldsymbol{\varphi}^h, \mathbb{p}^h; \delta\boldsymbol{\varphi}^h, \delta\mathbb{p}^h) = \int_0^L & (\delta\boldsymbol{\varphi}^h_{,s} \cdot \mathbf{E}(\mathbb{q}^h) + \frac{2}{\|\mathbb{p}^h\|}\boldsymbol{\varphi}^h_{,s} \cdot \mathbf{E}(\delta\mathbb{p}^h)) \cdot \mathbf{G}(\mathbb{q}^h)^\top \cdot \mathbf{N}^h \\ & - \delta\mathbb{p}^h \cdot \frac{2}{\|\mathbb{p}^h\|^2}\Big[(\mathbb{p}^h \otimes \mathbf{R}(\mathbb{q}^h)^\top \cdot \boldsymbol{\varphi}^h_{,s})\mathbf{N}^h \\ & + 2(\mathbb{q}^h \otimes \mathbf{G}(\mathbb{q}^h) \cdot \mathbb{p}^h_{,s}) \cdot \mathbf{M}^h\Big] \\ & + \frac{2}{\|\mathbb{p}^h\|^2}\left(\delta\mathbb{p}^h_{,s} \cdot \mathbf{G}(\mathbb{p}^h)^\top - \delta\mathbb{p}^h \cdot \mathbf{G}(\mathbb{p}^h_{,s})^\top\right) \cdot \mathbf{M}^h \, ds\,. \end{aligned} \tag{6.50}$$

Here, $\mathbf{N}^h = \mathbf{D}_1 \cdot \boldsymbol{\Gamma}^h$ and $\mathbf{M}^h = \mathbf{D}_2 \cdot \mathbf{K}^h$, where the discrete strain measures are given by Eq. (6.37) and Eq. (6.38). The last equation can also be written in the form

$$G^h_{\text{int}}(\boldsymbol{\varphi}_i, \mathbb{p}_i; \delta\boldsymbol{\varphi}_i, \delta\mathbb{p}_i) = \sum_{i=1}^n \begin{bmatrix} \delta\boldsymbol{\varphi}_i \\ \delta\mathbb{p}_i \end{bmatrix}^\top \int_0^L \begin{bmatrix} \mathbf{B}^{i\top}_{\varphi\varphi} & \mathbf{0} \\ \mathbf{B}^{i\top}_{\mathbb{p}\varphi} & \mathbf{B}^{i\top}_{\mathbb{p}\mathbb{p}} \end{bmatrix} \begin{bmatrix} \mathbf{N}^h \\ \mathbf{M}^h \end{bmatrix} ds\,, \tag{6.51}$$

where the nodal operator matrices are given by

$$\begin{aligned} \mathbf{B}^i_{\varphi\varphi} &= N_{i,s}\mathbf{R}(\mathbb{q}^h)^\top\,, \\ \mathbf{B}^i_{\mathbb{p}\varphi} &= \frac{2N_i}{\|\mathbb{p}^h\|^2}\left(\mathbf{G}(\mathbb{p}^h)^\top \cdot \overset{+}{\boldsymbol{\varphi}}{}^h_{,s} - \mathbf{R}(\mathbb{q}^h)^\top \cdot \boldsymbol{\varphi}^h_{,s} \otimes \mathbb{p}^h\right), \\ \mathbf{B}^i_{\mathbb{p}\mathbb{p}} &= \frac{2}{\|\mathbb{p}^h\|^2}\left[N_{i,s}\mathbf{G}(\mathbb{p}^h) - N_i\left(\mathbf{G}(\mathbb{p}^h_{,s}) + 2\mathbf{G}(\mathbb{q}^h) \cdot \mathbb{p}^h_{,s} \otimes \mathbb{q}^h\right)\right]. \end{aligned} \tag{6.52}$$

Note that in Eq. (6.52)$_2$ use has been made of the notation introduced in Eq. (6.8). An alternative form of Eq. (6.51) is given by

$$G_{\text{int}}^h(\boldsymbol{\varphi}_i, \mathbb{p}_i; \delta\boldsymbol{\varphi}_i, \delta\mathbb{p}_i) = \sum_{i=1}^{n} \left(\delta\boldsymbol{\varphi}_i \cdot \mathbf{f}_{\text{int}}^{\varphi,i} + \delta\mathbb{p}_i \cdot \mathbf{f}_{\text{int}}^{\mathbb{p},i} \right), \tag{6.53}$$

where

$$\begin{aligned} \mathbf{f}_{\text{int}}^{\varphi,i} &= \int_0^L \mathbf{B}_{\varphi\varphi}^{i^\top} \cdot \mathbf{N}^h \, \mathrm{d}s, \\ \mathbf{f}_{\text{int}}^{\mathbb{p},i} &= \int_0^L \mathbf{B}_{\mathbb{p}\varphi}^{i^\top} \cdot \mathbf{N}^h + \mathbf{B}_{\mathbb{p}\mathbb{p}}^{i^\top} \cdot \mathbf{M}^h \, \mathrm{d}s. \end{aligned} \tag{6.54}$$

denote the internal forces corresponding to the control/nodal points.

Remark 6.1. *Expression G_{int}^h in Eq. (6.50) can also be obtained by starting from*

$$G_{\text{int}}^h(\boldsymbol{\varphi}^h, \mathbb{p}^h; \delta\boldsymbol{\varphi}^h, \delta\mathbb{p}^h) = \int_0^L \delta\boldsymbol{\Gamma}^h \cdot \mathbf{N}^h + \delta\mathbf{K}^h \cdot \mathbf{M}^h \, \mathrm{d}s, \tag{6.55}$$

and taking into account the discrete strain measures Eq. (6.37) and Eq. (6.38).

6.4.2.1. Contribution of the external loading

The contribution of the external loading to the weak form follows from Eq. (6.28) by taking into account Eq. (6.32) and Eq. (6.35). Accordingly, we obtain

$$G_{\text{ext}}^h(\boldsymbol{\varphi}^h, \mathbb{p}^h; \delta\boldsymbol{\varphi}^h, \delta\mathbb{p}^h) = \int_0^L \delta\boldsymbol{\varphi}^h \cdot \bar{\mathbf{n}} + \delta\mathbb{p}^h \cdot \frac{2}{\|\mathbb{p}^h\|} \mathbf{E}(\mathbb{q}^h)^\top \cdot \bar{\mathbb{m}} \, \mathrm{d}s \tag{6.56}$$

leading to

$$G_{\text{ext}}^h(\boldsymbol{\varphi}_i, \mathbb{p}_i; \delta\boldsymbol{\varphi}_i, \delta\mathbb{p}_i) = \sum_{i=1}^{n} \left(\delta\boldsymbol{\varphi}_i \cdot \mathbf{f}_{\text{ext}}^{\varphi,i} + \delta\mathbb{p}_i \cdot \mathbf{f}_{\text{ext}}^{\mathbb{p},i} \right), \tag{6.57}$$

where

$$\begin{aligned} \mathbf{f}_{\text{ext}}^{\varphi,i} &= \int_0^L N_i \bar{\mathbf{n}} \, \mathrm{d}s \\ \mathbf{f}_{\text{ext}}^{\mathbb{p},i} &= \int_0^L \frac{2N_i}{\|\mathbb{p}^h\|} \mathbf{E}(\mathbb{q}^h)^\top \cdot \bar{\mathbb{m}} \, \mathrm{d}s \end{aligned} \tag{6.58}$$

denote the discrete external forces corresponding to the control/nodal points.

6.4.2.2. Contribution of the unit-length constraints

As outlined above we impose the unit-length constraint on the quaternions to resolve the redundancy of the quaternions at the control/nodal points. To this end we investigate two alternative procedures for imposing the unit-length constraint

$$g(\mathbb{p}) = \frac{1}{2}\left(\mathbb{p}\cdot\mathbb{p} - 1\right) = 0 \tag{6.59}$$

on the quaternion field. While the first option imposes the unit-length constraint in weak form, the second option is based on a strong (or point-wise) imposition at the control/nodal points.

Weak enforcement The weak imposition of the constraint Eq.(6.59) yields the following contribution of the conjugate constraint forces to the weak form

$$\begin{aligned} G_\lambda^h(\mathbb{p}^h, \boldsymbol{\lambda}^h; \delta\mathbb{p}^h) &= \int_0^L \delta g(\mathbb{p}^h)\lambda^h \mathrm{d}s \\ &= \int_0^L \delta\mathbb{p}^h \cdot \mathbb{p}^h \lambda^h \mathrm{d}s\,. \end{aligned} \tag{6.60}$$

In addition to that, the weak enforcement of the unit-length constraint is based on the condition

$$\int_0^L \delta\lambda^h g(\mathbb{p}^h)\mathrm{d}s = 0\,. \tag{6.61}$$

Concerning the discretization of the Lagrange multiplier field we apply a Bubnov-Galerkin approach which applies the same NURBS/Lagrangian shape functions as before. Accordingly,

$$\lambda^h(s) = \sum_{i=1}^n N_i(s)\lambda_i \quad \text{and} \quad \delta\lambda^h(s) = \sum_{i=1}^n N_i(s)\delta\lambda_i\,. \tag{6.62}$$

Making use of these ansatz functions in Eq.(6.61) and Eq.(6.60) yields the discrete constraint functions

$$g_i = \int_0^L N_i g(\mathbb{p}^h)\mathrm{d}s = 0 \tag{6.63}$$

for $i = 1, \ldots, n$, together with the contribution to the weak form of the discrete constraint forces

$$G_\lambda^h(\mathbb{p}_i, \boldsymbol{\lambda}_i; \delta\mathbb{p}_i) = \sum_{i=1}^n \delta\mathbb{p}_i \cdot \mathbf{f}_\lambda^i\,, \tag{6.64}$$

where the discrete constraint forces corresponding to the control/nodal points are given by

$$\mathbf{f}_\lambda^{\mathbb{p},i} = \int_0^L N_i \mathbb{p}^h \lambda^h \, \mathrm{d}s\,. \tag{6.65}$$

Strong enforcement To account for the redundancy of the quaternions $\mathbb{p}_i \in \mathbb{H}^4$, the unit-length constraint can also be enforced directly at the control/nodal points. In this case, the discrete constraints are given by

$$g_i = g(\mathbb{p}_i) = \frac{1}{2} (\mathbb{p}_i \cdot \mathbb{p}_i - 1) = 0 \tag{6.66}$$

for $i = 1, \ldots, n$. The contribution of the conjugate constraint forces to the weak form can again be written in the form Eq. (6.64), where

$$\mathbf{f}_\lambda^{\mathbb{p},i} = \mathbb{p}_i \lambda_i \tag{6.67}$$

denote the constraint forces corresponding to the control/nodal points.

6.4.3. Algebraic system of equations

To summarize the discretization procedure described above, weak form Eq.(6.49) eventually yields

$$\sum_{i=1}^{n} \begin{bmatrix} \delta\boldsymbol{\varphi}_i \\ \delta\mathbb{p}_i \\ \delta\lambda_i \end{bmatrix}^\top \left(\begin{bmatrix} \mathbf{f}_{\text{int}}^{\varphi,i} \\ \mathbf{f}_{\text{int}}^{\mathbb{p},i} \\ 0 \end{bmatrix} + \begin{bmatrix} \mathbf{0} \\ \mathbf{f}_\lambda^{\mathbb{p},i} \\ g_i \end{bmatrix} - \begin{bmatrix} \mathbf{f}_{\text{ext}}^{\varphi,i} \\ \mathbf{f}_{\text{ext}}^{\mathbb{p},i} \\ 0 \end{bmatrix} \right) = 0\,, \tag{6.68}$$

which has to hold for arbitrary $(\delta\boldsymbol{\varphi}_i, \delta\mathbb{p}_i, \delta\lambda_i) \in \mathbb{R}^3 \times \mathbb{H} \times \mathbb{R}$ $(i = 1, \ldots, n)$. One may either apply the strong or the weak enforcement of the constraints described in Section 6.4.2.2. To solve the algebraic system of nonlinear equations we apply Newton's method. Accordingly, in each Newton iteration a saddle point system has to be solved in order to eventually determine $(\boldsymbol{\varphi}_i, \mathbb{p}_i, \lambda_i) \in \mathbb{R}^3 \times \mathbb{H} \times \mathbb{R}$ $(i = 1, \ldots, n)$. In this connection a standard additive update procedure is applied in each Newton iteration.

6.4.3.1. Elimination of the Lagrange multipliers and the constraints

The strong enforcement of the constraints makes possible to apply a simple procedure for the elimination of the discrete constraint forces along with the constraints. This approach is referred to as the discrete null space method [18, 19, 92]. The discrete null space method essentially relies on two steps. Starting from the algebraic constraint Eq. (6.66), the corresponding consistency condition $\delta g_i = \mathbb{p}_i \cdot \delta\mathbb{p}_i = 0$ can be identically fulfilled by choosing $\delta\mathbb{p}_i = \mathbf{G}(\mathbb{p}_i)^\top \delta\boldsymbol{\theta}_i$ for any $\delta\boldsymbol{\theta}_i \in \mathbb{R}^3$. The last relationship for $\delta\mathbb{p}_i$ can be inserted into Eq. (6.68) to annihilate the constraint forces since

$$\left(\mathbf{G}(\mathbb{p}_i)^\top \cdot \delta\boldsymbol{\theta}_i\right) \cdot \mathbf{f}_\lambda^{\mathbb{p},i} = \delta\boldsymbol{\theta}_i \cdot \mathbf{G}(\mathbb{p}_i) \cdot \mathbf{f}_\lambda^{\mathbb{p},i} = \delta\boldsymbol{\theta}_i \cdot \mathbf{G}(\mathbb{p}_i) \cdot \mathbb{p}_i \lambda_i = 0$$

for arbitrary $\delta\boldsymbol{\theta}_i \in \mathbb{R}^3$ due to Eq. (6.6)$_1$. Accordingly, Eq. (6.68) can be recast in the size-reduced form

$$\sum_{i=1}^{n} \begin{bmatrix} \delta\boldsymbol{\varphi}_i \\ \delta\boldsymbol{\theta}_i \end{bmatrix}^{\mathsf{T}} \left(\begin{bmatrix} \mathbf{f}_{\text{int}}^{\varphi,i} \\ \mathbf{G}(\mathbb{p}_i) \cdot \mathbf{f}_{\text{int}}^{\mathbb{p},i} \end{bmatrix} - \begin{bmatrix} \mathbf{f}_{\text{ext}}^{\varphi,i} \\ \mathbf{G}(\mathbb{p}_i) \cdot \mathbf{f}_{\text{ext}}^{\mathbb{p},i} \end{bmatrix} \right) = 0\,, \tag{6.69}$$

which has to hold for arbitrary $(\delta\boldsymbol{\varphi}_i, \delta\boldsymbol{\theta}_i) \in \mathbb{R}^3 \times \mathbb{R}^3$ $(i = 1, \ldots, n)$. Of course, the constraints Eq. (6.66), i.e. $g_i = 0$, still have to be satisfied. However, in a second step we make these constraints obsolete by replacing the redundant unknowns $\mathbb{p}_i \in \mathbb{H}$ with new unknowns $\boldsymbol{\theta}_i \in \mathbb{R}^3$ such that

$$\mathbb{p}_i = \exp_{S^3}((0, \frac{1}{2}\boldsymbol{\Theta}_i)) \circ \mathbb{p}_i^0\,, \tag{6.70}$$

where $\exp_{S^3} : T_{\mathbb{p}_i^0} S^3 \mapsto S^3$ denotes the exponential map on S^3. Provided that the reference value $\mathbb{p}_i^0 \in S^3$, the update formula ensures $\mathbb{p}_i \in S^3$ so that the unit-length constraint is automatically satisfied. In equilibrium problems, $\mathbb{p}_i^0 \in S^3$ may be chosen to coincide with the last equilibrium configuration. Then, the application of a new load increment yields a new equilibrium configuration which is characterized by the rotation increments $\boldsymbol{\theta}_i \in \mathbb{R}^3$. During a load step the incremental rotations $\boldsymbol{\theta}_i$ are updated additively. Further details of the discrete null space method in the context of unit quaternions can be found in [20].

6.5. Numerical validation of the unit quaternion formulation

In this section we present several numerical examples to examine the numerical performance of alternative element formulations. In particular, we investigate to what extend the quaternion projection Eq. (6.29) does improve the performance when compared to the classical quaternion discretization, as discussed in Section 3.4.1.2, based on

$$\mathbb{p}^h(s) = \sum_{i=1}^{n} N_i(s)\mathbb{p}_i\,. \tag{6.71}$$

Furthermore, we impose the unit-length condition on the quaternions either in weak or in strong form, as described in Section 6.4.2.2. Last but not least we compare the NURBS-based formulation with that based on Lagrange shape functions.

To distinguish between the alternative formulations we introduce the abbreviations summarized in Tables 6.1 and 6.2. For example, NPS3 means the NURBS-based formulation of order three relying on the projection Eq. (6.29) and the strong enforcement of the unit-length constraint Eq. (6.66). Similarly, LCS2 means the Lagrange element of order two based on the classical approximation Eq. (6.71) and the strong enforcement of the unit-length constraint Eq. (6.66).

Table 6.1.: Abbreviations for elements based on NURBS shape functions [148].

shape functions	projection	constraints	order	abbreviation
			1	NPS1
NURBS	yes	strong	2	NPS2
			3	NPS3
			1	NPW1
NURBS	yes	weak	2	NPW2
			3	NPW3
			1	NCS1
NURBS	no	strong	2	NCS2
			3	NCS3
			1	NCW1
NURBS	no	weak	2	NCW2
			3	NCW3

Table 6.2.: Abbreviations for elements based on Lagrange shape functions [148].

shape functions	projection	constraints	order	abbreviation
			1	LPS1
Lagrange	yes	strong	2	LPS2
			3	LPS3
			1	LCS1
Lagrange	no	strong	2	LCS2
			3	LCS3

Concerning the initialization of the quaternions in the control points of the NURBS-based elements, we first compute auxiliary director frames at the Gauss points from which unit-quaternions can be extracted by applying Spurrier's algorithm [138]. The computation of the director frame is described in Section 5.6. Subsequently, the unit quaternions values of the control points are calculated by applying the procedure originally described in Section 2.4.2. Due to the interpolatory nature of the Lagrange shape functions, the initial director frames can be directly set up at the nodes. Application of Spurrier's algorithm directly yields the nodal unit quaternions.

To eliminate transverse shear locking we apply selectively reduced numerical integration. In particular, we choose the number of Gauss points equal to the element order.

The mesh refinement for the NURBS-based elements was performed with the algorithm proposed in [78], to keep the Jacobian of the isoparametric transformation constant over the domain. The mesh refinement for the Lagrange elements was performed by using an equidistant distribution of the finite element nodes.

The examples investigated in this chapter are the same as in the previous chapter (cf. Section 5.6). Therefore, we do not describe each problem itself anymore but focus on the discussion of the results.
A direct comparison of the convergence behavior to the results of the previous chapter is of limited use. Due to the classical discretization approach used in Chapter 5, the convergence behavior cannot be considered as optimal. Furthermore, as discussed in Section 6.5 the Gauss-Lobatto-based approach leads to a larger error for the first-order classical discretization approaches, which limits the relevance of a comparisons to the approaches presented in this chapter.
Thus, no such discussion is presented.

6.5.1. Cantilever beam under end moment

All details to the investigated problem, like geometrical and stiffness parameters as well as loading are given in Section 5.6.1.

Convergence plots for the cantilever beam under an end moment are displayed in Fig. 6.1. As in Section 5.6.1 the position of the tip of the beam structure is used as an error measure $e = ||\boldsymbol{\varphi}_{\text{num}}(s = L)||$, where $\boldsymbol{\varphi}_{\text{num}}(s = L)$ is the position vector of the tip of the numerical model. As in the previous chapter the load was applied in 10 equally spaced incremental steps. The convergence criteria of the Newton solver $||\mathbf{R}|| = 1 \times 10^{-12}$, where $||\mathbf{R}||$ is the 2-norm of the residual vector.

The convergence behavior is shown in Fig. 6.1 for various element formulations under consideration. In particular, Fig. 6.1a shows the results for NURBS shape functions of order $p = 1, \dots, 3$. Accordingly, the projection-based approach shows the best convergence behavior for both the strong enforcement of the unit-length constraint (NPSp, $p = 1, \dots, 3$) and the weak enforcement (NPWp, $p = 1, \dots, 3$). In both cases the rate of convergence turns out to be $\mathcal{O}(h^{2p})$.

However, if the projection is skipped the results worsen significantly if the strong enforcement of the unit-length constraint is used (Fig. 6.1a, NCSp, $p = 1, \dots, 3$). In particular, the order of convergence does not improve with increased order of the NURBS shape functions but stays at a level of $\mathcal{O}(h^2)$. This can be explained by the non-interpolatory nature of NURBS shape functions. Thus, enforcing the unit-length constraint at the control points in general does not imply unit-length of the quaternion discretization inside the computational domain. Still considering the classical interpolation (without projection) but choosing the weak enforcement instead of the strong enforcement of the unit-length constraint improves the convergence results (see NCWp, $p = 1, \dots, 3$, Fig. 6.1a).

It is also worth mentioning that the projection-based formulation in general exhibits a superior convergence behavior in the iterative solution procedure, in the sense that coarser meshes still reach a solution. In particular, while the projection-based formulation reaches a solution for the coarse discretization with 8 elements, this is not the case for the classical formulation which requires more than 8 elements to reach converged results (cf. Fig. 6.1a).

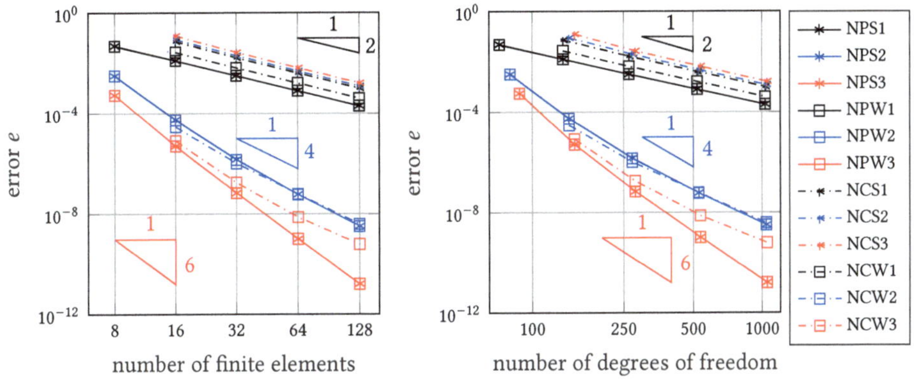

(a) Convergence behavior for the Isogeometric Analysis [148].

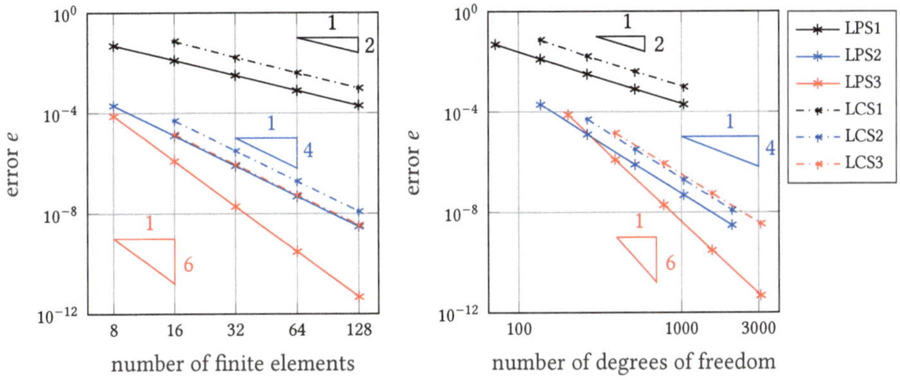

(b) Convergence behavior for the classical Lagrange elements.

Figure 6.1.: Convergence behavior of the quaternion beam formulation for the cantilever beam under end moment [148].

We now turn to the convergence results of the elements based on the Lagrange shape functions (Fig. 6.1b) where we focus our attention to the strong nodal enforcement of the unit-length constraint. Accordingly, the projection-based formulation (LPSp, $p = 1, \ldots, 3$) again shows superior convergence of order $\mathcal{O}(h^{2p})$, in analogy to the NURBS-based case. The classical formulation (LCSp, $p = 1, \ldots, 3$) again shows an order reduction, which, however, is not as pronounced as in the case of NURBS shape function. Specifically, for Lagrange elements the order reduction is deferred to $p \geq 3$. This slightly improved convergence behavior presumably is caused by the interpolatory nature of Lagrange polynomials.

As already shown in Chapter 2 in Fig. 2.11 the error bound per degree of freedom is lower for the NURBS basis function, when compared with the Lagrange approximation for the projection-based approach. This hold for all orders $p = 1, ..., 3$.

The deformed configurations of the beam corresponding to five different load levels are shown in Fig. 6.2. As expected, a complete roll-up takes place for $M = 2\pi \frac{EI}{L}$.

Fig. 6.2 displays the beam's configuration under different load levels during the roll-up process.

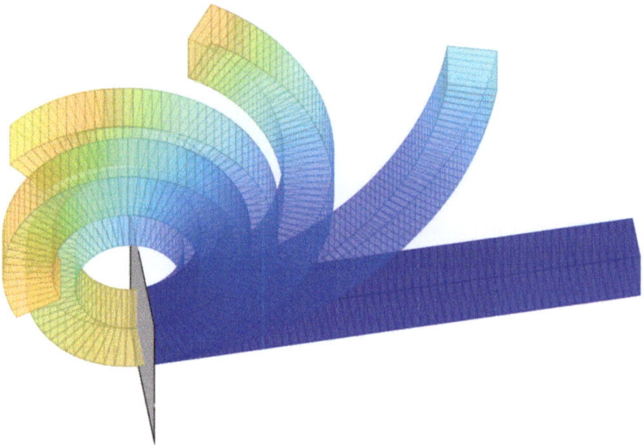

Figure 6.2.: Snapshots of the roll-up movement of the cantilever beam (32 NPS3 elements) for the load levels $M = 0, \frac{1}{5}M, \frac{2}{5}M, \frac{3}{5}M, \frac{4}{5}M, M = 2\pi \frac{EI}{L}$.

Additional results of the convergence behavior of Gauss-Lobatto spectrals elements are given in the Appendix C .

6.5.2. 3D bending and twist

A detailed description of the problem at hand can be found in the previous chapter (see Section 5.6.2)).

The corresponding formulation to apply an external moment in the body-fixed frame is given in the Appendix C
Fig. 6.4 shows three deformed configurations of the beam corresponding to three different load levels. In Fig. 6.3 the convergence behavior of the alternative beam elements under consideration is shown for a slenderness ratio of $\zeta = 25$. The error measure plotted is given by $e = \frac{\| \boldsymbol{\varphi}_{\text{num}}(s=L) - \boldsymbol{\varphi}_{\text{ana}}(s=L) \|}{\| \boldsymbol{\varphi}_{\text{num}}(s=L) \|}$, where $\boldsymbol{\varphi}_{\text{ana}}(s = L)$ is the analytical solution for the displacement of the tip under the final external torque $\bar{\mathbf{m}}$.

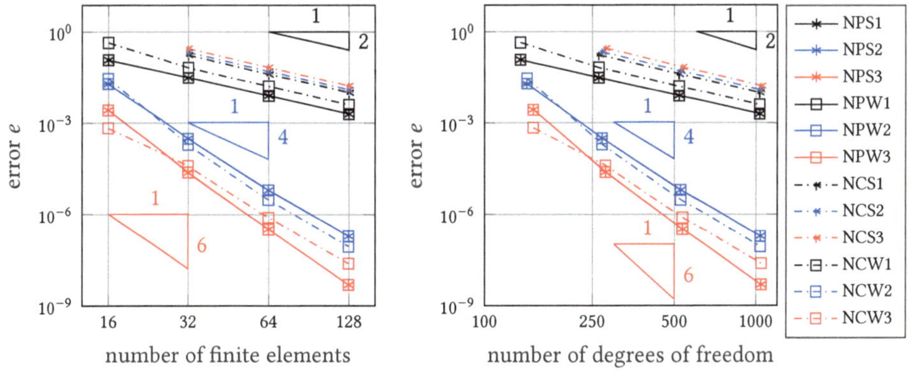

(a) Convergence behavior for the Isogeometric Analysis [148].

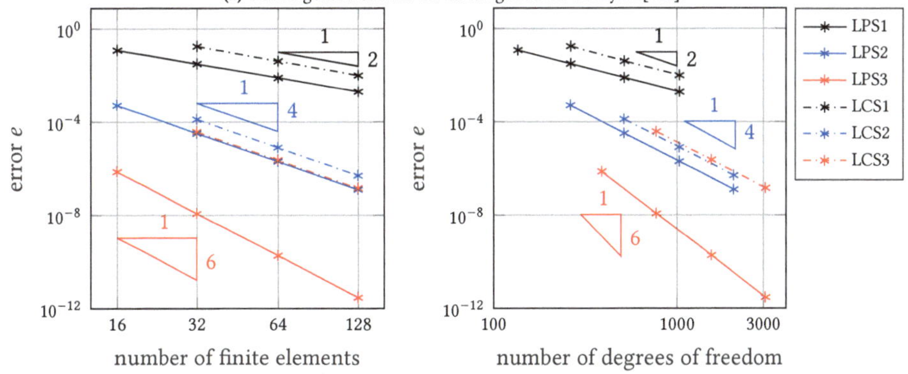

(b) Convergence behavior for the classical Lagrange elements.

Figure 6.3.: Convergence behavior of the quaternion beam formulation [148].

The convergence behavior is in complete analogy to that of the previous example. Accordingly, the projection-based approach yields superior convergence rates of $\mathcal{O}(h^{2p})$, together with a more robust solution behavior, in the sense that already 16 elements suffice to yield convergence of the iterative solution procedure. In contrast to that, the classical formulation (without projection) requires more elements to yield converged results. In addition to that, the classical formulation again exhibits order reduction for $p \geq 2$, which is especially pronounced for NCS2 and NCS3.

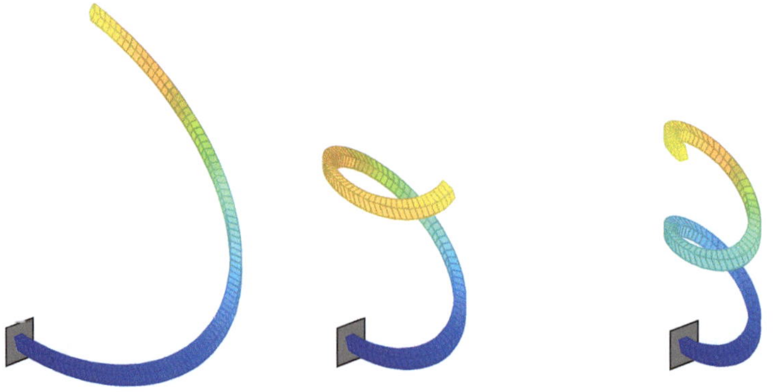

(a) Load level: $\frac{8}{25}\bar{\mathbf{m}}$ - 80 loads steps. **(b)** Load level: $\frac{16}{25}\bar{\mathbf{m}}$ - 160 load steps. **(c)** Load level: $\bar{\mathbf{m}}$ - 250 load steps.

Figure 6.4.: Configurations of the deformed beam corresponding to different numbers of load steps. Eventually, a helix with two complete coils is reached [148].

Gauss-Lobatto spectral elements additional results showing the convergence behavior are given in the Appendix C.

6.5.3. Numerical path-independence test

In the previous chapter (Section 5.6.3) all details, such as loading and geometrical parameters, to this example are given.

The cantilever bend is discretized with 32 linear elements. The convergence criteria for the Netwon solver is set to $||\mathbf{R}|| = 1 \times 10^{-6}$. In Tab. 6.3 the displacement of the tip in the direction of $\mathbf{e}_2$ is presented. At the end of the cycle the load is removed, so that the end configuration has to coincide with the initial configuration.
All presented results are in good agreement with the results presented in [21]. The end configuration coincides again with the initial configuration up to an error due to the Newton method for the presented quaternion formulation. In contrast the path-dependent '$\Delta\theta$' proposed in [127, 128] does not return to the initial configuration.
Additional results for different numbers of elements ([8, 16, 32]) and orders $p = 1, ..., 3$ are given in the Appendix C.

Table 6.3.: Numerical path-independence test: tip displacement in 2-direction corresponding to specific load levels [148].

load level	tip displacement in 2-direction		
	$\Delta\theta$ formulation [21]	Lagrange (LPS1)	NURBS (NPS1)
$[F_1,\ F_2,\ F_3]$	32 el. $p = 1$	32 el. $p = 1$	32 el. $p = 1$
$[0\ \ 0\ \ 0]$	0	0	0
$[-600\ \ 0\ \ 0]$	0	-8.8844×10^{-20}	-2.3202×10^{-14}
$[-600\ \ 600\ \ 0]$	59.7884	59.8262	59.8796
$[-600\ \ 600\ \ 600]$	38.6655	38.6875	38.6952
$[0\ \ 600\ \ 600]$	37.5087	37.5269	37.5247
$[0\ \ 0\ \ 600]$	0.0190	-1.6362×10^{-11}	-1.6369×10^{-11}
$[0\ \ 0\ \ 0]$	0.0374	1.2632×10^{-23}	-7.8410×10^{-16}

In Tab. 6.4 the convergence behavior of the Newton solver for eight linear elements are shown for the last loading step. In comparison to the director formulation (see. Tab. 5.3) the quaternion formulation leads to a convergence in less steps. It can be observed that in general the quaternion formulation shows a much more robust convergence behavior of the Newton solver than the director formulation.

Table 6.4.: Residual of Newton solver for last time step with 8 linear elements.

Iteration number	LPS1 2-norm of residual	NPS1 2-norm of residual
1	2.50×10^{1}	2.50×10^{1}
2	3.14×10^{4}	8.93×10^{4}
3	4.11×10^{1}	3.86×10^{1}
4	5.81×10^{1}	1.35×10^{2}
5	2.76×10^{-5}	1.40×10^{-3}
6	2.60×10^{-8}	7.00×10^{-7}

6.5.4. Beam patches with slope discontinuity

In Section 5.6.4 all details to the problem can be found. We limit ourselves here to presenting the obtained results.

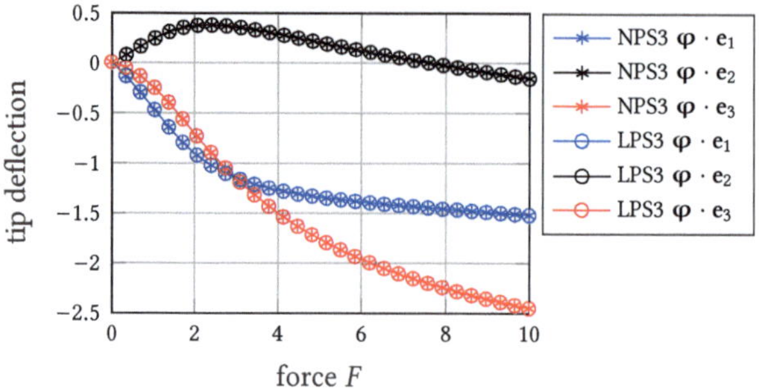

Figure 6.5.: Deflection of the tip versus force [148].

In Fig. 6.5 the components of the tip deflection are plotted over the load level. Both NURBS (NPS3) and Lagrange (LPS3) elements are used with a strong enforcement of the unit-length constraint. It can be seen that the results of the NURBS and Lagrange elements agree very well. In particular, no difference can be observed in Fig. 6.5. Furthermore, the results are in very good agreement with those in [45, 65, 115].

The configuration for different load levels is shown in Fig. 6.6. As in the previous chapter the load is applied in 30 equally space steps. In Fig. 6.6a the configuration is shown after 10 load steps, in Fig. 6.6b after 20 load steps and in Fig. 6.6c the whole load of $\mathbf{F} = \begin{bmatrix} -10 & 0 - 10 \end{bmatrix}^{\mathsf{T}}$ is applied.

The formulation of a rigid joint in a quaternion setting is given in the Appendix C.

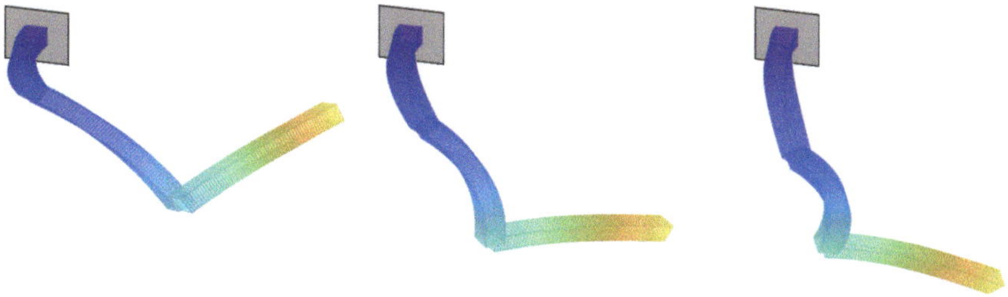

(a) Load level: $\frac{1}{3}$F - 10 load steps. (b) Load level: $\frac{2}{3}$F - 20 load steps. (c) Load level: F - 30 load steps.

Figure 6.6.: Configuration of the beam patches for different load levels [148].

6.5.5. Conclusion

A quaternion formulation represents a very good alternative to the director formulation. The continuous formulation is not convoluted, while it allows for a simple discretization based on the projection-based elements. Due to its simplicity the projection-based approach is efficient. It leads to a path-independent and frame-indifferent formulation, which can be utilized with Lagrange or NURBS functions without any further adaptations and results in an optimal convergence behavior for all orders. Furthermore, it allows for an arbitrary continuity of the underlying discretization over element boundaries in the IGA.

7. Conclusion

7.1. Summary

The work investigates possible parametrization of rotations and their appropriate treatment within the discretization in the FEM and IGA, where the focus lies on an application in the context of the geometrically exact beam.

For this purpose, at the beginning, the fundamentals for the Isogeometric Analysis are recaptured. Various aspects of the method, such as refinement strategies, and numerical integration are discussed in detail. Furthermore, several advantages over the FEM, are presented.
In a subsequent chapter, rotations in three dimensions and possible parametrization of them are discussed. For a broader understanding selected topics of the underlying Lie group theory are presented. Furthermore, a discussion of possible discretization approaches for rotational variables is given. Hereby, lies the focus on the applicability in the context of FEM and IGA as well as the feasibility of the approaches in applications.
The geometrically exact beam is introduced beginning with a literature survey of the topic in the next chapter. The beam kinematics are presented and the equations of motions are derived by employing the balance equations. Furthermore, the kinetic and inner potential together with the constitutive law are presented. A very brief overview of the challenges of the shear constrained Kirchhoff-Love beam is given.
Chapter 5 presents in detail the geometrically exact beam in a director formulation. First, the continuous weak form is derived and in a subsequent step, the finite element formulation with an additive approach is presented. Two approaches to enforce the orthonormality constraints, necessary in this formulation, are introduced. Furthermore, the applied energy-momentum conserving time integration scheme based on the discrete gradient is presented. The convergence behavior with respect to mesh-refinement and order elevation is discussed for Lagrange, NURBS, and Gauss-Lobatto-based shape functions. Furthermore, the path-independence of the formulation is shown in a numerical simulation. At last, dynamic examples are presented for a discretization with Lagrange and NURBS elements.
A formulation of the geometrically exact beam using unit quaternions is presented in Chapter 6. At the beginning of the chapter, the continuous beam equations are presented in a unit quaternion setting. The frame-indifference of the corresponding continuous strain measures is demonstrated and the associated weak form of the quaternion formulation is derived. In the following, the resulting equations are discretized using a

projection-based discretization approach, which conserves the underlying geometry of S^3. To handle the redundancy of the discretized equations, two approaches to eliminate it and enforce the unit constraints are discussed, which can be applied to NURBS and Lagrange shape functions. Excellent results for IGA and FEM are obtained for static examples, where the convergence behavior is presented and discussed in detail. Furthermore, an example shows, that the formulation leads to a path-independent formulation.

7.2. Conclusion

Nowadays the IGA can be considered a well-established simulation method in the scientific community. As discussed in Section 2.4.4, the IGA offers many advantages over the standard FEM. It is, thus, desirable to be able to use the method for all kinds of problems. However, so far the existing literature is very limited with respect to the treatment of the geometrically exact beam model in an isogeometric framework. This work aims to close this gap.

The main difficulty, when dealing with a beam model in a Galerkin method, such as the FEM and IGA, is the correct treatment of rotations. As many examples from the literature show, a frame-indifferent and path-independent discretized model is not always given. A basic knowledge of the underlying mathematical structure, the Lie group theory, is useful to construct appropriate discretization approaches. Furthermore, a discretization approach needs to represent the underlying geometry of the rotational variable sufficiently to achieve convergence behavior for order elevation. As examples from this work show, this is only given if the discretization is performed with great care. Groundbreaking in this regard are the concepts of the geodesic and projection-based finite elements, as they show a structured approach to constructing conforming elements of arbitrary order for the FEM and IGA. Furthermore, they are proven to result in optimal error bounds [58, 59, 120] and result by construction in path-independent and frame-indifferent formulations [59].
We discussed the discretization for a director framework and unit quaternions to represent rotations. Hereby, it clearly shows, that an appropriate choice to represent rotations is essential since it simplifies the discretization approach greatly.
While the director approach leads to mathematical simple expressions in the continuous case, the correct discretization is sophisticated and computationally expensive. This holds for geodesic as well as projection-based elements.

The far better choice to parametrize rotations is the use of unit quaternions. While quaternions themselves, are an advanced concept with a new algebra, their discretization approach in a geodesic formulation has the same level of complexity as for directors. However, the projection-based approach introduced in Chapter 3 leads to a discretization of unit quaternions with a relatively simple approach, which is computationally efficient. The projection-based approach leads not just to the correct preservation of the unit sphere, but it also results in a model, where the derivatives and variations lie in the correct

tangent space. Furthermore, the increase in the dimension compared to the minimal set is moderate.
In addition, in the context of the beam, it is possible to perform a strong enforcement of the unit length constraints in the isogeometric framework. As discussed in Section 6.4.3.1, this allows the elimination of constraints and a formulation on the minimal set of coordinates. Research needs to be conducted, if this can be generalized for all projection-based element formulations in combination with the IGA.
Furthermore, in the boundary value problems investigated in this work, the two-to-one representation of $SO(3)$ given by quaternions does not cause any issues. Compared to the investigated director formulation, the quaternion formulation results in a more robust model, which is less prone to locking.
However, due to the projection-based approach and its introduced nonlinearities, the design of a time integration scheme, which conserves energy, as well as linear and angular momentum, is not straightforward and could pose a challenging task.

Also, for the FEM the use of Gauss-Lobatto-based shape functions in combination with the Lobatto integration is an interesting approach due to its simplicity. The implementation is straightforward and no further adaptations of the finite element code are needed. Even though this approach is greatly limited due to the fixed combination of integration points and nodes, it could pose a simple solution for individual problems or to verify implementations of complex models, before implementing more convoluted discretization approaches.

7.3. Outlook

This work gives an extensive overview of the treatment of rotations in a finite element and isogeometric framework. Of course, this does not imply that research in this regard is completed, as many more aspects still are to be investigated.
While the geodesic and projection-based discretization approaches for the directors are presented in theory in Chapter 3, they are not applied in this work. Even though these approaches are not very well suited for a practical application due to the associated high computational cost, it still would be of interest to compare the results of the element formulations for the FEM as well as the IGA in detail. A detailed discussion of higher order geodesic elements, even though they were already presented in 1999 [33] is not given in the literature for the IGA to the best knowledge of the author. One focus of interest for both element formulations is hereby the enforcement of the orthogonality constraints in the case of NURBS functions. Advantageous would be a strong enforcement as it would allow for the elimination of the constraints using the null space method (cf. Section 5.5.1).

As mentioned above, the best-suited parametrization in the FEM and IGA for rotations seems to be the use of unit quaternions due to the simplicity of the projection-based formulation. The projection-based elements offer a relatively simple formulation for unit quaternions and the formulation can be applied to either FEM or IGA without any further

adaptations. Due to their simplicity, they are computationally efficient. However, the literature reports a lower error bound for the geodesic elements [59, 102], so a direct comparison of both approaches would be of interest. To the best knowledge of the other, such a discussion has so far not been presented in the literature.
Furthermore, the unit quaternion formulation could be investigated in connection with time integration methods. Investigations in combination with structure-preserving time integrator schemes are so far missing in the literature. In the first step, it might be promising to design a structure-preserving integrator without the projection-based approach to simplify the problem. In the second step, the projection could be integrated into the scheme.
The obtained knowledge could, further, be utilized to design structure-preserving integrators for the geometrically exact beam formulated in dual quaternions. Dual quaternions seem to be a promising parametrization for the beam, since like an $SE(3)$ formulation, a finite element beam formulation in dual quaternions is free of locking [91].
Another interesting utilization of the projection-based approach of unit quaternions could be in a $SE(3)$ beam formulation. Here the quaternions could be used to parametrize the tensors of $SO(3)$. A projection-based approach for the special Euclidean group itself might lead to a very complicated approach as already the projection-based approach for the $SO(3)$ is convoluted. Due to its high computational cost, a geodesic approach for the $SE(3)$ formulation is limited for practical applications.

Furthermore, the projection-based approach for unit quaternions could be used in Galerkin-based time integration schemes, for the geometrically exact beam but likewise for rigid body mechanics. A similar approach could also be applied for Galerkin-based integrators, to problems such as the Reissner-Mindlin shell.

A. Appendix to Chapter 3

A.1. Projector onto the tangent space

The orthogonal mappings onto the tangent space $\mathbf{P}_{T\mathcal{M}} : \mathbb{R}^n \mapsto T_{\mathbf{x}}\mathcal{M}$ presented in Section 3.4.3 are projectors [2]. Thus, they fulfill the requirement of idempotency

$$\mathbf{P}_{T\mathcal{M}} = \mathbf{P}_{T\mathcal{M}} \cdot \mathbf{P}_{T\mathcal{M}} . \tag{A.1}$$

Their counterpart, the orthogonal projector onto the normal space of $\mathcal{M}$ can be simply be derived from the completeness condition, such that the projector onto the normal space follows from

$$\mathbf{P}_{T\mathcal{M}}^{\perp} = \mathbf{I} - \mathbf{P}_{T\mathcal{M}} . \tag{A.2}$$

Both projectors are orthogonal

$$\mathbf{P}_{T\mathcal{M}} \cdot \mathbf{P}_{T\mathcal{M}}^{\perp} = \mathbf{0} . \tag{A.3}$$

The orthogonal projector onto the normal space of $\mathcal{M}$ is also idempotent

$$\mathbf{P}_{T\mathcal{M}}^{\perp} = \mathbf{P}_{T\mathcal{M}}^{\perp} \cdot \mathbf{P}_{T\mathcal{M}}^{\perp} . \tag{A.4}$$

B. Appendix to Chapter 5

B.1. Numerical path-independence test

Additional results for the approximation with 8, 16 and 32 finite elements of element orders $p = 1, \ldots, 3$ are presented as mentioned in 5.6.3. The number of elements is based on the results given in [21], where only results for first-order elements given. Gauss-Lobatto-based shape functions with a Lobatto integration leading to very strong locking effects for first order elements, thus, for these elements only results from higher-order elements are presented as discussed in 5.6.3.
The presented values are evaluated at the same load levels as in Tab. 5.2.

Table B.1.: Numerical path-independence test: tip displacement $\boldsymbol{\varphi} \cdot \mathbf{e}_2$ corresponding to specific load levels - additional results for different number of elements and linear elements ($p = 1$).

	tip displacement $\boldsymbol{\varphi} \cdot \mathbf{e}_2$ Lagrange elements (LCS1)	
8 el.	16 el.	32 el.
0	0	0
0	0	0
61.3302	60.1164	59.9022
40.0323	38.9324	38.7539
38.3769	37.7249	37.5829
1.8201×10^{-17}	2.6430×10^{-17}	2.8962×10^{-17}
$\ll \varepsilon$	$\ll \varepsilon$	$\ll \varepsilon$

Table B.2.: Numerical path-independence test: tip displacement $\boldsymbol{\varphi} \cdot \mathbf{e}_2$ corresponding to specific load levels - additional results for different number of elements and linear elements ($p = 1$).

tip displacement $\boldsymbol{\varphi} \cdot \mathbf{e}_2$ NURBS elements (NCW1)		
8 el.	16 el.	32 el.
0	0	0
0	0	0
59.6563	59.915	59.8467
38.4339	38.7477	38.7036
37.2041	37.5625	37.5391
1.2323×10^{-17}	1.2275×10^{-17}	1.2474×10^{-17}
$\ll \varepsilon$	$\ll \varepsilon$	$\ll \varepsilon$

Table B.3.: Numerical path-independence test: tip displacement $\boldsymbol{\varphi} \cdot \mathbf{e}_2$ corresponding to specific load levels - additional results for different number of elements and quadratic elements ($p = 2$).

tip displacement $\boldsymbol{\varphi} \cdot \mathbf{e}_2$ Lagrange elements (LCS2)		
8 el.	16 el.	32 el.
0	0	0
0	0	0
59.8482	59.8346	59.8338
38.7106	38.6982	38.6974
37.5427	37.5368	37.5364
2.9282×10^{-17}	2.9407×10^{-17}	2.9694×10^{-17}
$\ll \varepsilon$	$\ll \varepsilon$	$\ll \varepsilon$

Table B.4.: Numerical path-independence test: tip displacement $\boldsymbol{\varphi} \cdot \mathbf{e}_2$ corresponding to specific load levels - additional results for different number of elements and quadratic elements ($p = 2$).

tip displacement $\boldsymbol{\varphi} \cdot \mathbf{e}_2$ NURBS elements (NCW2)		
8 el.	16 el.	32 el.
0	0	0
0	0	0
59.7388	59.8718	59.8755
38.5969	38.7335	38.7375
37.4546	37.5782	37.5810
1.2557×10^{-17}	1.2814×10^{-17}	1.2546×10^{-17}
$\ll \varepsilon$	$\ll \varepsilon$	$\ll \varepsilon$

Table B.5.: Numerical path-independence test: tip displacement $\boldsymbol{\varphi} \cdot \mathbf{e}_2$ corresponding to specific load levels - additional results for different number of elements and quadratic elements (p = 2).

tip displacement $\boldsymbol{\varphi} \cdot \mathbf{e}_2$ Gauss-Lobatto-based elements (GCS2)		
8 el.	16 el.	32 el.
0	0	0
0	0	0
57.2126	59.5662	59.8105
37.0237	38.4395	38.6740
34.4074	37.1419	37.5062
2.5813×10^{-17}	3.0828×10^{-17}	3.0172×10^{-17}
$\ll \varepsilon$	$\ll \varepsilon$	$\ll \varepsilon$

Table B.6.: Numerical path-independence test: tip displacement $\boldsymbol{\varphi} \cdot \mathbf{e}_2$ corresponding to specific load levels - additional results for different number of elements and cubic elements (p = 3).

tip displacement $\boldsymbol{\varphi} \cdot \mathbf{e}_2$ Lagrange elements (LCS3)		
8 el.	16 el.	32 el.
0	0	0
0	0	0
59.8359	59.8339	59.8338
38.6993	38.6975	38.6974
37.5374	37.5364	37.5364
2.9687×10^{-17}	2.8628×10^{-17}	3.1478×10^{-17}
$\ll \varepsilon$	$\ll \varepsilon$	$\ll \varepsilon$

Table B.7.: Numerical path-independence test: tip displacement $\boldsymbol{\varphi} \cdot \mathbf{e}_2$ corresponding to specific load levels - additional results for different number of elements and cubic elements (p = 3).

tip displacement $\boldsymbol{\varphi} \cdot \mathbf{e}_2$ NURBS elements (NCW3)		
8 el.	16 el.	32 el.
0	0	0
0	0	0
59.7962	59.8736	59.8755
38.6689	38.7358	38.7375
37.5397	37.5801	37.5810
1.2796×10^{-17}	1.2731×10^{-17}	1.2596×10^{-17}
$\ll \varepsilon$	$\ll \varepsilon$	$\ll \varepsilon$

Table B.8.: Numerical path-independence test: tip displacement $\boldsymbol{\varphi} \cdot \mathbf{e}_2$ corresponding to specific load levels - additional results for different number of elements and cubic elements ($p = 3$).

tip displacement $\boldsymbol{\varphi} \cdot \mathbf{e}_2$ Gauss-Lobatto-based elements (GCS3)		
8 el.	16 el.	32 el.
0	0	0
0	0	0
59.7891	59.8326	59.8338
38.6562	38.6962	38.6974
37.5070	37.5357	37.5364
2.9427×10^{-17}	3.0390×10^{-17}	3.7597×10^{-26}
$\ll \varepsilon$	$\ll \varepsilon$	$\ll \varepsilon$

C. Appendix to Chapter 6

Appendix C.1 to Appendix C.3 and Appendix C.4.2 are taken from [148]. Minor adaptations were made to fit into the framework of this work.

C.1. Strain measure for bending and twist

We verify that strain measure Eq. (6.17) corresponds to the axial vector of the skew-symmetric matrix Eq. (6.16). Accordingly, one has to show that $\mathbf{K} = 2\mathbf{G}(\mathbb{q}) \cdot \mathbb{q}_{,s}$ is the axial vector of matrix $2\mathbf{G}(\mathbb{q}) \cdot \mathbf{G}(\mathbb{q}_{,s})^\top$. To see this, consider the identity

$$\mathbf{G}(\mathbb{q})\mathbb{q}_{,s} = -q_{0,s}\mathbf{q} + q_0\mathbf{q}_{,s} - \mathbf{q} \times \mathbf{q}_{,s} \,. \tag{C.1}$$

The skew-symmetric matrix associated with the above vector is given by

$$\begin{aligned} \mathbf{S} &= -q_{0,s}\widehat{\mathbf{q}} + q_0\widehat{\mathbf{q}}_{,s} - \widehat{\mathbf{q} \times \mathbf{q}_{,s}} \\ &= -q_{0,s}\widehat{\mathbf{q}} + q_0\widehat{\mathbf{q}}_{,s} - \left(\mathbf{q}_{,s} \otimes \mathbf{q} - \mathbf{q} \otimes \mathbf{q}_{,s}\right) . \end{aligned} \tag{C.2}$$

Since $\widehat{\mathbf{a}}\widehat{\mathbf{b}} = \mathbf{b} \otimes \mathbf{a} - (\mathbf{a} \cdot \mathbf{b}) \cdot \mathbf{I}_3$ for any $\mathbf{a}, \mathbf{b} \in \mathbb{R}^3$,

$$\mathbf{q}_{,s} \otimes \mathbf{q} = \widehat{\mathbf{q}} \cdot \widehat{\mathbf{q}}_{,s} + (\mathbf{q}_{,s} \cdot \mathbf{q}) \cdot \mathbf{I}_3 \,. \tag{C.3}$$

Moreover, since $\mathbb{q} \in S^3$ implies $\mathbb{q}_{,s} \cdot \mathbb{q} = 0$ or $\mathbf{q}_{,s} \cdot \mathbf{q} = -q_{0,s}q_0$, we eventually obtain

$$\mathbf{S} = -q_{0,s}\widehat{\mathbf{q}} + q_0\widehat{\mathbf{q}}_{,s} + \mathbf{q} \otimes \mathbf{q}_{,s} - \widehat{\mathbf{q}} \cdot \widehat{\mathbf{q}}_{,s} + q_0 q_{0,s}\mathbf{I}_3 \,. \tag{C.4}$$

A direct calculation yields the result $\mathbf{G}(\mathbb{q}) \cdot \mathbf{G}(\mathbb{q}_{,s})^\top = \mathbf{S}$.

C.2. Application of external torques

In the examples dealt with in Sections 6.5.1 and 6.5.2 an external torque $\bar{\mathbf{m}} \in \mathbb{R}^3$ is applied at one end of the beam. Since the collocation property holds at the end-points of both finite element formulations under consideration, the contribution to the weak form is given by

$$\begin{aligned} G^e_{\text{ext}} &= \delta \mathbb{q}_e \cdot 2\mathbf{E}(\mathbb{q}_e)^\top \cdot \bar{\mathbf{m}} \\ &= \delta \mathbb{p}_e \cdot \frac{1}{\|\mathbb{p}_e\|} (\mathbf{I}_4 - \mathbb{q}_e \otimes \mathbb{q}_e) \cdot 2\mathbf{E}(\mathbb{q}_e)^\top \cdot \bar{\mathbf{m}} \\ &= \delta \mathbb{p}_e \cdot \frac{2}{\|\mathbb{p}_e\|^2} \mathbf{E}(\mathbb{p}_e)^\top \cdot \bar{\mathbf{m}} , \end{aligned} \tag{C.5}$$

where the quaternion normalization procedure has been accounted for. Furthermore, index e refers to the control/nodal point at the end of the beam. While in Section 6.5.1 the external torque $\bar{\mathbf{m}} = M\mathbf{e}_3$ can be directly inserted into the above formula, the external torque in Section 6.5.2 is a follower load defined by $\bar{\mathbf{m}} = \bar{m}^i \mathbf{d}^e_i = \mathbf{E}(\mathbb{q}_e) \cdot \mathbf{G}(\mathbb{q}_e)^\top \cdot \bar{m}^i \mathbf{e}_i$. Inserting this expression into the above formula yields

$$G^e_{\text{ext}} = \delta \mathbb{p}_e \cdot \frac{2}{\|\mathbb{p}_e\|^2} \mathbf{G}(\mathbb{p}_e)^\top \cdot \bar{m}^i \mathbf{e}_i . \tag{C.6}$$

Here, the normalization procedure along with the properties summarized in Eq. (6.6) have been taken into account.

C.3. Rigid intersection between two beams for the quaternion formulation

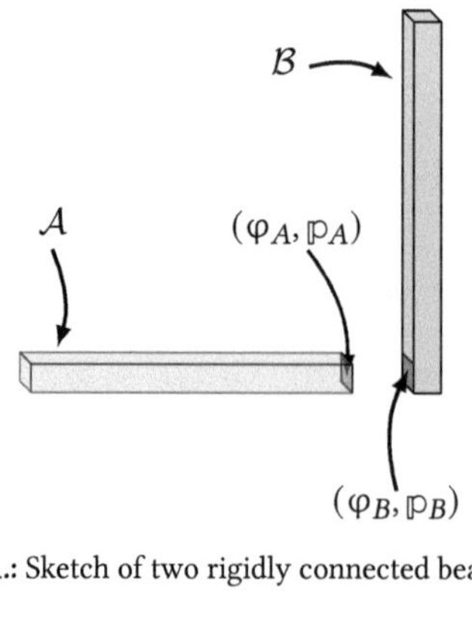

Figure C.1.: Sketch of two rigidly connected beams $\mathcal{A}$ and $\mathcal{B}$.

In this Appendix we address the rigid connection of two beams as required in the numerical example in Section 6.5.4. In particular, we consider a rigid intersection between the end-points of two beams (Fig. C.1). Accordingly, the two end-points, say A and B, experience the same rotation $\mathbb{r} \in S^3$ with respect to some previous cross-sectional orientations

$\mathbb{q}_{\mathrm{A}}^{0} \in S^3$ and $\mathbb{q}_{\mathrm{B}}^{0} \in S^3$. That is, the new cross-sectional orientations are given by $\mathbb{p}_{\mathrm{A}} = \mathbb{r} \circ \mathbb{q}_{\mathrm{A}}^{0}$ and $\mathbb{p}_{\mathrm{B}} = \mathbb{r} \circ \mathbb{q}_{\mathrm{B}}^{0}$. These two equations can be rewritten as $\mathbb{r} = \mathbb{p}_{\mathrm{A}} \circ \bar{\mathbb{q}}_{\mathrm{A}}^{0}$ and $\mathbb{r} = \mathbb{p}_{\mathrm{B}} \circ \bar{\mathbb{q}}_{\mathrm{B}}^{0}$, respectively. Since $\mathbb{r} \circ \bar{\mathbb{r}} = (1, \mathbf{0})$, we obtain

$$\mathbb{p}_{\mathrm{A}} \circ \bar{\mathbb{q}}_{\mathrm{A}}^{0} \circ \mathbb{q}_{\mathrm{B}}^{0} \circ \bar{\mathbb{p}}_{\mathrm{B}} = (1, \mathbf{0}) \,. \tag{C.7}$$

We take the vector part of the above condition to obtain three equations of constraint to impose the rigid intersection between the two beams. Three more conditions arise from the additional constraint $\boldsymbol{\varphi}_{\mathrm{A}} - \boldsymbol{\varphi}_{\mathrm{B}} = \mathbf{0}$. Accordingly, the rigid coupling of two beams can be realized by appending a total of six algebraic constraints to the beam formulation in Section 6.4.3.

Alternatively, algebraic constraints can be circumvented by applying the discrete null space method outlined in Section 6.4.3.1. Accordingly, starting with the constrained formulation in Section 6.4.3, the quantities $(\boldsymbol{\varphi}_{\mathrm{B}}, \mathbb{p}_{\mathrm{B}})$ and $(\delta\boldsymbol{\varphi}_{\mathrm{B}}, \delta\mathbb{p}_{\mathrm{B}})$ related to control/nodal point B can be expressed in terms of the corresponding quantities in point A via

$$\boldsymbol{\varphi}_{\mathrm{B}} = \boldsymbol{\varphi}_{\mathrm{A}} \,, \tag{C.8}$$

$$\mathbb{p}_{\mathrm{B}} = \mathbb{p}_{\mathrm{A}} \circ \bar{\mathbb{q}}_{\mathrm{A}}^{0} \circ \mathbb{q}_{\mathrm{B}}^{0} \,, \tag{C.9}$$

and

$$\delta\boldsymbol{\varphi}_{\mathrm{B}} = \delta\boldsymbol{\varphi}_{\mathrm{A}} \,, \tag{C.10}$$

$$\delta\mathbb{p}_{\mathrm{B}} = \delta\mathbb{p}_{\mathrm{A}} \circ \bar{\mathbb{q}}_{\mathrm{A}}^{0} \circ \mathbb{q}_{\mathrm{B}}^{0} \,. \tag{C.11}$$

A further size-reduction of the algebraic system to be solved can be achieved by applying the procedure described in Section 6.4.3.1.

C.4. Additional numerical results

We present here additional results accommodating the results presented in Chapter 6. Tab. C.1 introduces the abbreviations used in the following in combination with the quaternion formulation introduced in Section 6.5.

Table C.1.: Abbreviations for elements based on Gauss-Lobatto-based shape functions.

shape functions	projection	constraints	order	abbreviation
			1	GCS1
Gauss-Lobatto-based	no	strong	2	GCS2
			3	GCS3

C.4.1. Cantilever under end moment

The results presented here are obtained from the problem described in Section 5.6.1. We present here the convergence behavior of the GCSp elements alongside results form Section 6.5.1 obtained with the LPSp elements for direct comparison. Fig. C.2 depicts the convergence behavior for both type of elements. It can be seen that the GCSp elements show the same order of convergence of $\mathcal{O}(h^{2p})$ as the LPSp elements. However, for all orders the GCSp elements show a larger error, which is in accordance to the results presented in Chapter 5.

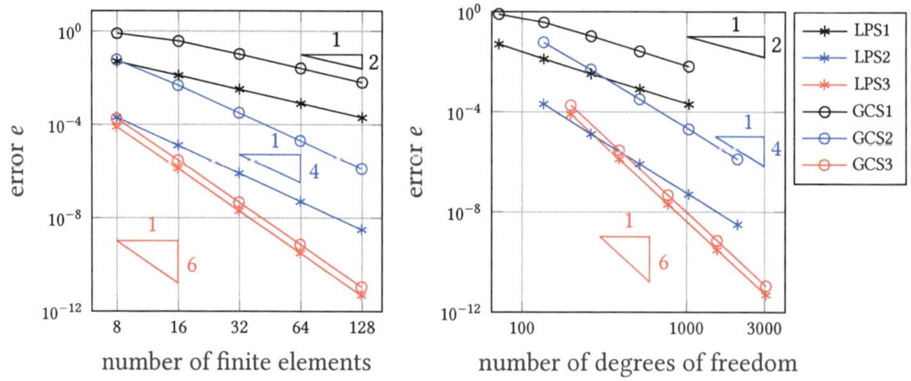

Figure C.2.: Convergence behavior of the quaternion beam formulation for the roll-up into a circle.

C.4.2. 3D bending and twist

The problem, for which the results are shown here, is described in detail in Section 5.6.2. As before the results of the GSCp elements are presented together with the results from the LPSp elements from Section 6.5.2 in Fig. C.3. The GSCp elements for $p = 1, 2$ show

the same convergence order of $\mathcal{O}(h^{2p})$ as the LPSp elements. However, the results for the GSC3 elements show a very larger error, with approximately the same convergence order of $\mathcal{O}(h^{2})$ as the GSC1 or LPS1 elements. The obtained error is even slightly bigger per degree of freedom compared to the GSC1 elements.
It is not obvious, why this phenomena does not occur for the previous example presented in Fig. C.2.

C.4.3. Numerical path-independence test

As mentioned in Section 6.5.3 we here show the results for other discretizations with 8, 16 and 32 elements and element order p = 1, ..., 3. The number of elements is analogous

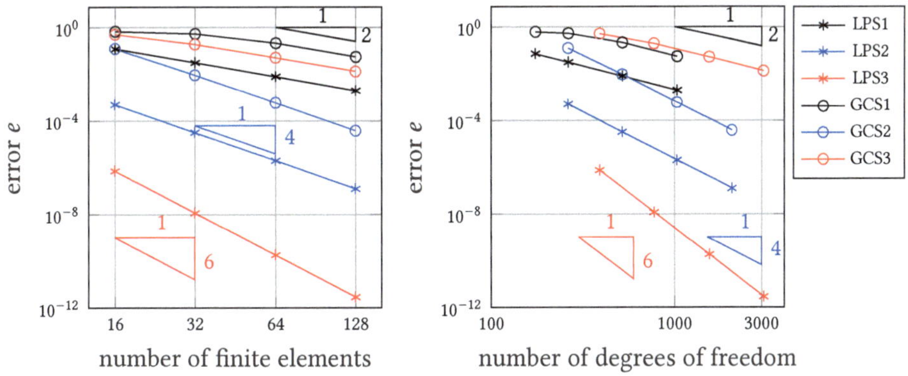

Figure C.3.: Convergence behavior of the quaternion beam formulation for the roll-up into a helix.

to the results found in [21], where only results for a first order discretization are shown. As we use NURBS basis functions we are able to conserve the geometry of the 1/8-circle exactly for the order p = 2, 3. For this purpose the initial control points of the NURBS formulation and their weights were computed with an algorithm from Chapter 7.5 in [110] to define a section of a circle in combination with the refinement strategy for k-refinement in Chapter 5.5 in [110] and for h-refinement in [78] as mentionend in Section 6.5.
All values are evaluated for the same loads as in Tab. 6.3.

Table C.2.: Numerical path-independence test: tip displacement $\boldsymbol{\varphi} \cdot \mathbf{e}_2$ corresponding to specific load levels - additional results for different number of elements and linear elements ($p = 1$).

	tip displacement $\boldsymbol{\varphi} \cdot \mathbf{e}_2$ Lagrange elements (LPS1)	
8 el.	16 el.	32 el.
0	0	0
-8.9232×10^{-20}	-8.6515×10^{-20}	-8.844×10^{-20}
59.7206	59.8038	59.8262
38.5488	38.6585	38.6875
37.3891	37.4987	37.5269
-1.5671×10^{-11}	-1.6225×10^{-11}	-1.6362×10^{-11}
4.1576×10^{-22}	-4.9593×10^{-23}	1.2632×10^{-23}

Table C.3.: Numerical path-independence test: tip displacement $\boldsymbol{\varphi} \cdot \mathbf{e}_2$ corresponding to specific load levels - additional results for different number of elements and linear elements ($p = 1$).

tip displacement $\boldsymbol{\varphi} \cdot \mathbf{e}_2$ NURBS elements (NPS1)		
8 el.	16 el.	32 el.
0	0	0
1.2277×10^{-14}	-1.1828×10^{-15}	-2.3202×10^{-14}
59.7132	59.8029	59.8261
38.5469	38.6584	38.6875
37.3916	37.4491	37.5270
-1.5783×10^{-11}	-1.6246×10^{-11}	-1.6369×10^{-11}
-1.1852×10^{-15}	9.2804×10^{-16}	-7.8410×10^{-16}

Table C.4.: Numerical path-independence test: tip displacement $\boldsymbol{\varphi} \cdot \mathbf{e}_2$ corresponding to specific load levels - additional results for different number of elements and quadratic elements ($p = 2$).

tip displacement $\boldsymbol{\varphi} \cdot \mathbf{e}_2$ Lagrange elements (LPS2)		
8 el.	16 el.	32 el.
0	0	0
1.8588×10^{-17}	6.4369×10^{-18}	9.3133×10^{-21}
58.6129	59.7042	59.8232
37.7556	38.5692	38.6869
35.7855	37.3584	37.5233
-1.1916×10^{-11}	-1.6153×10^{-11}	-1.6386×10^{-11}
-7.4184×10^{-20}	-1.002×10^{-21}	8.9024×10^{-22}

Table C.5.: Numerical path-independence test: tip displacement $\boldsymbol{\varphi} \cdot \mathbf{e}_2$ corresponding to specific load levels - additional results for different number of elements and quadratic elements ($p = 2$).

tip displacement $\boldsymbol{\varphi} \cdot \mathbf{e}_2$ NURBS elements (NPS2)		
8 el.	16 el.	32 el.
0	0	0
-4.9258×10^{-14}	-1.1828×10^{-15}	-2.3202×10^{-14}
57.6927	59.5842	59.8542
37.1466	38.4948	38.7195
35.1708	37.3045	37.5619
-8.6269×10^{-12}	-1.5806×10^{-11}	-1.6403×10^{-11}
5.0916×10^{-14}	6.6981×10^{-15}	-1.4206×10^{-14}

Table C.6.: Numerical path-independence test: tip displacement $\boldsymbol{\varphi} \cdot \mathbf{e}_2$ corresponding to specific load levels - additional results for different number of elements and cubic elements ($p = 3$).

tip displacement $\boldsymbol{\varphi} \cdot \mathbf{e}_2$ Lagrange elements (LPS3)		
8 el.	16 el.	32 el.
0	0	0
-2.7663×10^{-16}	5.9239×10^{-20}	-1.275×10^{-19}
59.8085	59.8332	59.8338
38.6732	38.6969	38.6974
37.5207	37.5361	37.5364
-1.6403×10^{-11}	-1.6408×10^{-11}	-1.6407×10^{-11}
-1.9518×10^{-21}	-2.0485×10^{-23}	-2.2070×10^{-12}

Table C.7.: Numerical path-independence test: tip displacement $\boldsymbol{\varphi} \cdot \mathbf{e}_2$ corresponding to specific load levels - additional results for different number of elements and cubic elements ($p = 3$).

tip displacement $\boldsymbol{\varphi} \cdot \mathbf{e}_2$ NURBS elements (NPS3)		
8 el.	16 el.	32 el.
0	0	0
-2.8256×10^{-14}	-1.0551×10^{-14}	-8.5072×10^{-15}
59.7312	59.8724	59.8754
38.6098	38.7347	38.7374
37.5061	37.5795	37.5809
-1.6504×10^{-11}	-1.6453×10^{-11}	-1.6448×10^{-11}
4.3234×10^{-16}	2.6724×10^{-15}	-2.5779×10^{-15}

D. Copyright statement

Publications and talks

Peer-reviewed publications

- Wasmer P and Betsch P. "A projection-based quaternion discretization of the geometrically exact beam model". In: *Int J Numer Meth Eng.* e7538, DOI: 10.1002/nme.7538.

Other publications

- Wasmer P and Betsch P. "Non-Linear Beam Formulation with NURBS Interpolation for the Simulation of Sliding Contacts". In: *ECCOMAS Thematic Conference on Multibody Dynamics.* 2021, pp. 136–147. DOI: 10.3311/ECCOMASMBD2021-181.
- Wasmer P and Betsch P. "Director-based IGA beam elements for sliding contact problems". In: *ECCOMAS Congress 2022-8th European Congress on Computational Methods in Applied Sciences and Engineering.* 2022.
- Wasmer P and Betsch P. "The geometrically exact beam model with a normalized quaternion discretization". In: *Proceedings of the 11th ECCOMAS Thematic Conference on Multibody Dynamics Lisbon, Portugal.* 2023.

Conference talks

- Wasmer P and Betsch P. *Non-Linear Beam Formulation with NURBS Interpolation for the Simulation of Sliding Contacts.* 10th ECCOMAS Thematic Conference on Multibody Dynamics, Budapest, Hungary, Dec. 13 – 15. 2021.
- Wasmer P and Betsch P. *Director-based IGA beam elements for sliding contact problems.* 8th European Congress on Computational Methods in Applied Sciences and Engineering (ECCOMAS), Oslo, Norway, June 5 - 9. 2022.
- Wasmer P and Betsch P. *Simulation of the geometrically exact beam model using Isogeometric Analysis.* 92nd Annual Meeting of the International Association of Applied Mathematics and Mechanics (GAMM), Aachen, Germany, Aug. 15 - 19. 2022.
- Wasmer P and Betsch P. *The geometrically exact beam model with a projected quaternion discretization.* 11th ECCOMAS Thematic Conference on Multibody Dynamics, Lisbon, Portugal, July 15 - 19. 2023.

- Wasmer P and Betsch P. *The geometrically exact beam with a projection-based discretization for unit quaternions.* 10th GACM Colloquium on Computational Mechanics for Young Scientists from Academia and Industry, Vienna, Austria, Sep. 10 - 13. 2023.
- Wasmer P and Betsch P. *The geometrically exact beam in a quaternion formulation with an energy-momentum conserving integrator.* 16th World Congress on Computational Mechanics and 4th Pan American Congress on Computational Mechanics, Vancouver, Canada, July 21 – 26. 2024.

List of Figures

List of Tables

Bibliography

[1] Absil PA, Mahony R, and Sepulchre R. *Optimization algorithms on matrix manifolds.* Princeton University Press, 2008.

[2] Absil PA, Mahony R, and Trumpf J. "An Extrinsic Look at the Riemannian Hessian". In: *Geometric Science of Information.* Ed. by Nielsen F and Barbaresco F. Springer, 2013, pp. 361–368. DOI: 10.1007/978-3-642-40020-9_39.

[3] Adam C, Bouabdallah S, Zarroug M, and Maitournam H. "Improved numerical integration for locking treatment in isogeometric structural elements, Part I: Beams". In: *Comput Methods Appl Mech Eng.* 279: 1–28, 2014. DOI: 10.1016/j.cma.2014.06.023.

[4] Antman SS. *Nonlinear Problems of Elasticity.* Springer-Verlag New York, 2005.

[5] Antman SS, Carlson DE, Fichera G, Grutin ME, and Naghdi PM. *Mechanics of Solids: Linear theories of elasticity and thermoelasticity. Linear and nonlinear theories of rods, plates, and shells.* Ed. by Truesdell C. Springer, 1984.

[6] Arnold VI. *Mathematical methods of classical mechanics.* Vol. 60. Springer Science & Business Media, 1989.

[7] Auricchio F, Calabrò F, Hughes TJR, Reali A, and Sangalli G. "A simple algorithm for obtaining nearly optimal quadrature rules for NURBS-based isogeometric analysis". In: *Comput Methods Appl Mech Eng.* 249-252: 15–27, 2012. DOI: 10.1016/j.cma.2012.04.014.

[8] Auricchio F, Carotenuto P, and Reali A. "On the geometrically exact beam model: A consistent, effective and simple derivation from three-dimensional finite-elasticity". In: *Int J Solids Struct.* 45(17): 4766–4781, 2008. DOI: 10.1016/j.ijsolstr.2008.04.015.

[9] Bartoň M and Calo VM. "Gauss–Galerkin quadrature rules for quadratic and cubic spline spaces and their application to isogeometric analysis". In: *Comput Aided Des.* 82: 57–67, 2017. DOI: 10.1016/j.cad.2016.07.003. eprint: 1602.01200.

[10] Bauchau O. *Flexible multibody dynamics.* Vol. 176. Springer, 2011.

[11] Bauchau O and Han S. "Interpolation of rotation and motion". In: *Multibody Syst Dyn.* 31: 339–370, 2014. DOI: 10.1007/s11044-013-9365-8.

[12] Bauer A, Breitenberger M, Philipp B, Wüchner R, and Bletzinger KU. "Nonlinear isogeometric spatial Bernoulli beam". In: *Comput Method App M.* 303: 101–127, 2016. DOI: 10.1016/j.cma.2015.12.027.

[13] Bazilevs Y, Beirão Da Veiga L, Cottrell JA, Hughes TJR, and Sangalli G. "Isogeometric analysis: Approximation, stability and error estimates for h-refined meshes". In: *Math Mod Meth Appl S.* 16(7): 1031–1090, 2006. DOI: 10.1142/S0218202506001455.

[14] Belytschko T, Liu WK, Moran B, and Elkhodary K. *Nonlinear finite elements for continua and structures.* John Wiley & Sons, 2014.

[15] Benzi M, Golub G, and Liesen J. "Numerical solution of saddle point problems". In: *Acta numer.* 14: 1–137, 2005. DOI: 10.1017/S0962492904000212.

[16] Bertram A. *Elasticity and plasticity of large deformations.* Springer, 2012.

[17] Betsch P and Sänger N. "On the consistent formulation of torques in a rotationless framework for multibody dynamics". In: *Comput Struct.* 127: 29–38, 2013. DOI: 10.1016/j.compstruc.2012.10.005.

[18] Betsch P. "The discrete null space method for the energy consistent integration of constrained mechanical systems. Part I: Holonomic constraints". In: *Comput Method App M.* 194(50-52): 5159–5190, 2005. DOI: 10.1016/j.cma.2005.01.004.

[19] Betsch P and Leyendecker S. "The discrete null space method for the energy consistent integration of constrained mechanical systems. Part II: Multibody dynamics". In: *Int J Numer Meth Eng.* 67(4): 499–552, 2006. DOI: 10.1002/nme.1639.

[20] Betsch P and Siebert R. "Rigid body dynamics in terms of quaternions: Hamiltonian formulation and conserving numerical integration". In: *Int J Numer Meth Eng.* 79(4): 444–473, 2009. DOI: 10.1002/nme.2586.

[21] Betsch P and Steinmann P. "Frame-indifferent beam finite elements based upon the geometrically exact beam theory". In: *Int J Numer Meth Eng.* 54(12): 1775–1788, 2002. DOI: 10.1002/nme.487.

[22] Betsch P and Steinmann P. "Constrained dynamics of geometrically exact beams". In: *Computat Mech.* 31: 49–59, 2003. DOI: 10.1007/s00466-002-0392-1.

[23] Betsch P and Uhlar S. "Energy-momentum conserving integration of multibody dynamics". In: *Multibody Syst Dyn.* 17: 243–289, 2007. DOI: 10.1007/s11044-007-9043-9.

[24] Bishop RL. "There is more than one way to frame a curve". In: *Am Math Mon.* 82(3): 246–251, 1975. DOI: 10.2307/2319846.

[25] Bosten A, Cosimo A, Linn J, and Brüls O. "A mortar formulation for frictionless line-to-line beam contact". In: *Multibody Syst Dyn.* 54(1): 31–52, 2022.

[26] Brüls O, Cardona A, and Arnold M. "Lie group generalized-α time integration of constrained flexible multibody systems". In: *Mech Mach Theory.* 48: 121–137, 2012. DOI: 10.1016/j.mechmachtheory.2011.07.017.

[27] Cardona A and Géradin M. "A beam finite element non-linear theory with finite rotations". In: *Int J Numer Meth Eng.* 26(11): 2403–2438, 1988. DOI: 10.1002/nme.1620261105.

[28] Celledoni E and Owren B. “Lie group methods for rigid body dynamics and time integration on manifolds”. In: *Comput Method Appl M.* 192(3-4): 421–438, 2003. DOI: 10.1016/s0045-7825(02)00520-0.

[29] Chen YC and Wheeler L. “Derivatives of the stretch and rotation tensors”. In: *J Elasticity.* 32: 175–182, 1993. DOI: 10.1007/bf00131659.

[30] Cottanceau E, Thomas O, Véron P, Alochet M, and Deligny R. “A finite element/quaternion/asymptotic numerical method for the 3D simulation of flexible cables”. In: *Finite Elem Anal Des.* 139: 14–34, 2018. DOI: 10.1016/j.finel.2017.10.002.

[31] Cottrell JA, Hughes TJR, and Reali A. “Studies of refinement and continuity in isogeometric structural analysis”. In: *Comput Methods Appl Mech Eng.* 196(41-44): 4160–4183, 2007. DOI: 10.1016/j.cma.2007.04.007.

[32] Cottrell JA, Hughes TJR, and Bazilevs Y. *Isogeometric Analysis - Toward Integration of CAD and FEA.* West Sussex: John Wiley & Sons, 2009. DOI: 10.1002/9780470749081.

[33] Crisfield MA and Jelenić G. “Objectivity of strain measures in the geometrically exact three-dimensional beam theory and its finite-element implementation”. In: *Proc R Soc A.* 455: 1125–1147, 1999. DOI: 10.1098/rspa.1999.0352.

[34] Curtis ML. *Matrix groups.* Springer New York, 1984.

[35] Da Veiga L., Buffa A, Sangalli G, and Vázquez R. “Mathematical analysis of variational isogeometric methods”. In: *Acta Numer.* 23: 157–287, 2014. DOI: 10.1017/s096249291400004x.

[36] Dam E, Koch M, and Lillholm M. *Quaternions, interpolation and animation.* Tech. rep. DIKU-TR-98/5. Department of Computer Science, University of Copenhagen, 1998. URL: https://web.mit.edu/2.998/www/QuaternionReport1.pdf.

[37] De Lorenzis L, Wriggers P, and Hughes TJR. “Isogeometric contact: a review”. In: *GAMM-Mitteilungen.* 37(1): 85–123, 2014. DOI: 10.1002/gamm.201410005.

[38] Debeurre M, Grolet A, and Thomas O. “Extreme nonlinear dynamics of cantilever beams: effect of gravity and slenderness on the nonlinear modes”. In: *Nonlinear Dynam.* 111(14): 12787–12815, 2023. DOI: 10.1007/s11071-023-08637-x.

[39] Demoures F, Gay-Balmaz F, Leyendecker S, Ober-Blöbaum S, Ratiu TS, and Weinand Y. “Discrete variational Lie group formulation of geometrically exact beam dynamics”. In: *Numer Math.* 130: 73–123, 2015. DOI: 10.1007/s00211-014-0659-4.

[40] Dichmann D, Li Y, and Maddocks J. “Hamiltonian Formulations and Symmetries in Rod Mechanics”. In: *Mathematical Approaches to Biomolecular Structure and Dynamics, IMA Volumes in Mathematics and its Applications.* Vol. 82. 1996, pp. 71–113. DOI: 10.1007/978-1-4612-4066-2_6.

[41] Dornisch W, Klinkel S, and Simeon B. “Isogeometric Reissner-Mindlin shell analysis with exactly calculated director vectors”. In: *Comput Method App M.* 253: 491–504, 2013. DOI: 10.1016/j.cma.2012.09.010.

[42] Echter R and Bischoff M. "Numerical efficiency, locking and unlocking of NURBS finite elements". In: *Comput Methods Appl Mech Eng.* 199(5-8): 374–382, 2010. DOI: 10.1016/j.cma.2009.02.035.

[43] Eisenberg M and Guy R. "A proof of the Hairy Ball Theorem". In: *Am Math Mon.* 86(7): 571–574, 1979. DOI: 10.2307/2320587.

[44] Eugster SR. *Geometric continuum mechanics and induced beam theories.* Springer, 2015.

[45] Eugster SR, Hesch C, Betsch P, and Glocker C. "Director-based beam finite elements relying on the geometrically exact beam theory formulated in skew coordinates". In: *Int J Numer Meth Eng.* 97(2): 111–129, 2014. DOI: 10.1002/nme.4586.

[46] Eugster SR and Harsch J. "A Variational Formulation of Classical Nonlinear Beam Theories". In: *Developments and Novel Approaches in Nonlinear Solid Body Mechanics.* Ed. by Abali BE and Giorgio I. Cham: Springer International Publishing, 2020, pp. 95–121. DOI: 10.1007/978-3-030-50460-1_9.

[47] Evans JA, Bazilevs Y, Babuška I, and Hughes TJR. "n-Widths, sup–infs, and optimality ratios for the k-version of the isogeometric finite element method". In: *Comput Method Appl M.* 198(21-26): 1726–1741, 2009. DOI: 10.1016/j.cma.2009.01.021.

[48] Fischer P, Klassen M, Mergheim J, Steinmann P, and Müller R. "Isogeometric analysis of 2D gradient elasticity". In: *Comput Mech.* 47: 325–334, 2011. DOI: 10.1007/s00466-010-0543-8.

[49] Gawlik E and Leok M. "Embedding-based interpolation on the special orthogonal group". In: *SIAM J Sci Comput.* 40(2): A721–A746, 2018. DOI: 10.1137/17m1129416.

[50] Gebhardt CG and Romero I. "On a nonlinear rod exhibiting only axial and bending deformations: mathematical modeling and numerical implementation". In: *Acta Mech.* 3825—3847: 625–647, 2021. DOI: 10.1007/s00707-021-03038-w.

[51] Ghosh S and Roy D. "Consistent quaternion interpolation for objective finite element approximation of geometrically exact beam". In: *Comput Method App M.* 198(3-4): 555–571, 2008. DOI: 10.1016/j.cma.2008.09.004.

[52] Gomez H, Hughes TJR, Nogueira X, and Calo VM. "Isogeometric analysis of the isothermal Navier–Stokes–Korteweg equations". In: *Comput Methods Appl Mech Eng.* 199(25-28): 1828–1840, 2010. DOI: 10.1016/j.cma.2010.02.010.

[53] Gonzalez O. "Time Integration and Discrete Hamiltonian Systems". In: *J Nonlinear Sci.* 6(5): 449–467, 1996. DOI: 10.1007/BF02440162.

[54] Gonzalez O. "Mechanical systems subject to holonomic constraints: Differential–algebraic formulations and conservative integration". In: *Physica D.* 132(1): 165–174, 1999. DOI: 10.1016/S0167-2789(99)00054-8.

[55] Gravenkamp H. "Numerical methods for the simulation of ultrasonic guided waves". Band 116. PhD thesis. Technische Universität Braunschweig, 2014. URL: https://opus4.kobv.de/opus4-bam/frontdoor/index/index/year/2015/docId/42.

[56] Greco L and Cuomo M. “B-Spline interpolation of Kirchhoff-Love space rods”. In: *Comput Method App M.* 256: 251–269, 2013. DOI: 10.1016/j.cma.2012.11.017.

[57] Greco L and Cuomo M. “An implicit G1 multi patch B-spline interpolation for Kirchhoff–Love space rod”. In: *Comput Method App M.* 269: 173–197, 2014. DOI: 10.1016/j.cma.2013.09.018.

[58] Grohs P, Hardering H, and Sander O. “Optimal A Priori Discretization Error Bounds for Geodesic Finite Elements”. In: *Found Comput Math.* 15(6): 1357–1411, 2015. DOI: 10.1007/s10208-014-9230-z.

[59] Grohs P, Hardering H, Sander O, and Sprecher M. “Projection-Based Finite Elements for Nonlinear Function Spaces”. In: *SIAM J Numer Anal.* 57(1): 404–428, 2019. DOI: 10.1137/18M1176798.

[60] Grohs P. “Quasi-interpolation in Riemannian manifolds”. In: *IMA J Numer Anal.* 33(3): 849–874, 2013. DOI: 10.1093/imanum/drs026.

[61] Hall BC. *Lie groups, Lie algebras, and representations.* Springer, 2013.

[62] Han S and Bauchau O. “On the global interpolation of motion”. In: *Comput Methods Appl Mech Eng.* 337: 352–386, 2018. DOI: 10.1016/j.cma.2018.04.002.

[63] Han S and Bauchau OA. “Configurational forces in variable-length beams for flexible multibody dynamics”. In: *Multibody Syst Dyn.* 58(3): 275–298, 2023. DOI: 10.21203/rs.3.rs-2035380/v1.

[64] Hanson AJ. *Visualizing quaternions.* Elsevier, 2005.

[65] Harsch J, Capobianco G, and Eugster SR. “Finite element formulations for constrained spatial nonlinear beam theories”. In: *Math Mech Solids.* 26(23): 1838–1863, 2021. DOI: 10.1177/10812865211000790.

[66] Harsch J, Sailer S, and Eugster S. “A total Lagrangian, objective and intrinsically locking-free Petrov–Galerkin SE (3) Cosserat rod finite element formulation”. In: *Int J Numer Meth Eng.* 124(13): 2965–2994, 2023. DOI: 10.1002/nme.7236.

[67] Harsch J and Eugster SR. “Finite Element Analysis of Planar Nonlinear Classical Beam Theories”. In: *Developments and Novel Approaches in Nonlinear Solid Body Mechanics.* Ed. by Abali BE and Giorgio I. Cham: Springer International Publishing, 2020, pp. 123–157. DOI: 10.1007/978-3-030-50460-1_10.

[68] Heaviside O. *Electromagnetic Theory - Volume I.* "The Electrian" Printing and Publishing Co. Limited, 1893.

[69] Hesch C, Schuß S, Dittmann M, Franke M, and Weinberg K. “Isogeometric analysis and hierarchical refinement for higher-order phase-field models”. In: *Comput Methods Appl Mech Eng.* 303: 185–207, 2016. DOI: 10.1016/j.cma.2016.01.022.

[70] Hiemstra RR, Calabrò F, Schillinger D, and Hughes TJR. “Optimal and reduced quadrature rules for tensor product and hierarchically refined splines in isogeometric analysis”. In: *Comput Methods Appl Mech Eng.* 316: 966–1004, 2017. DOI: 10.1016/j.cma.2016.10.049.

[71] Hilgert J and Neeb KH. *Lie-Gruppen und Lie-Algebren.* Springer-Verlag, 2013.

[72] Holm DD, Schmah T, and Stoica C. *Geometric Mechanics and Symmetry: From Finite to Infinite Dimensions.* Vol. 12. Oxford University Press, 2009.

[73] Holm DD. *Geometric mechanics-part II: rotating, translating and rolling.* World Scientific, 2011.

[74] Howe R. "Very basic Lie theory". In: *Am Math Mon.* 90(9): 600–623, 1983.

[75] Huang D and Leyendecker S. "An electromechanically coupled beam model for dielectric elastomer actuators". In: *Computat Mech.* 69(3): 805–824, 2022.

[76] Huang D and Leyendecker S. "Optimal control of dielectric elastomer actuated multibody dynamical systems". In: *Soft Robot.* 10(5): 897–911, 2023. DOI: 10.1007/s00466-021-02115-0.

[77] Hughes TJR. *The finite element method: linear static and dynamic finite element analysis.* Prentice-Hall, Inc., 1987.

[78] Hughes TJR, Cottrell J, and Bazilevs Y. "Isogeometric analysis: CAD, finite elements, NURBS, exact geometry and mesh refinement". In: *Comput Methods Appl Mech Eng.* 194(39): 4135–4195, 2005. DOI: 10.1016/j.cma.2004.10.008.

[79] Hughes TJR, Reali A, and Sangalli G. "Efficient quadrature for NURBS-based isogeometric analysis". In: *Comput Methods Appl Mech Eng.* 199(5-8): 301–313, 2010. DOI: 10.1016/j.cma.2008.12.004.

[80] Hughes TJR, Reali A, and Sangalli G. "Duality and unified analysis of discrete approximations in structural dynamics and wave propagation: comparison of p-method finite elements with k-method NURBS". In: *Comput Methods Appl Mech Eng.* 197(49-50): 4104–4124, 2008. DOI: 10.1016/j.cma.2008.04.006.

[81] Ibrahimbegović A. "On finite element implementation of geometrically nonlinear Reissner's beam theory: three-dimensional curved beam elements". In: *Comput Methods Appl Mech Eng.* 122(1-2): 11–26, 1995. DOI: 10.1016/0045-7825(95)00724-F.

[82] Iserles A, Munthe-Kaas HZ, Nørsett SP, and Zanna A. "Lie-group methods". In: *Acta Numer.* 1–148, 2005. DOI: 10.1017/s0962492900002154.

[83] Jelenić G and Crisfield M. "Geometrically exact 3D beam theory: implementation of a strain-invariant finite element for statics and dynamics". In: *Computer methods in applied mechanics and engineering.* 171(1-2): 141–171, 1999. DOI: 10.1016/s0045-7825(98)00249-7.

[84] Johannessen KA. "Optimal quadrature for univariate and tensor product splines". In: *Comput Methods Appl Mech Eng.* 316: 84–99, 2017. DOI: 10.1016/j.cma.2016.04.030.

[85] Karlin S and Schumaker L. "The fundamental theorem of algebra for Tchebycheffian monosplines". In: *J Anal Math.* 20(1): 233–270, 1967. DOI: 10.1007/BF02786674.

[86] Kiendl J, Bletzinger KU, Linhard J, and Wüchner R. "Isogeometric shell analysis with Kirchhoff–Love elements". In: *Comput Methods Appl Mech Eng.* 198(49-52): 3902–3914, 2009. DOI: 10.1016/j.cma.2009.08.013.

[87] Kim MJ, Kim MS, and Shin SY. "A C^2-continuous B-spline quaternion curve interpolating a given sequence of solid orientations". In: *Proceedings Computer Animation'95.* IEEE. 1995, pp. 72–81. DOI: 10.1109/ca.1995.393545.

[88] Kim MJ, Kim MS, and Shin SY. "A General Construction Scheme for Unit Quaternion Curves with Simple High Order Derivatives". In: *SIGGRAPH '95: Proceedings of the 22nd annual conference on Computer graphics and interactive techniques.* 1995, pp. 369–376. DOI: 10.1145/218380.218486.

[89] Knapp A. *Lie groups beyond an introduction.* Vol. 140. Springer, 1996.

[90] Lang H, Linn J, and Arnold M. "Multi-body dynamics simulation of geometrically exact Cosserat rods". In: *Multibody Syst Dyn.* 25(3): 285–312, 2011. DOI: 10.1007/s11044-010-9223-x.

[91] Leitz T and Leyendecker S. "Galerkin Lie-group variational integrators based on unit quaternion interpolation". In: *Comput Method Appl M.* 338: 333–361, 2018. DOI: 10.1016/j.cma.2018.04.022.

[92] Leyendecker S, Betsch P, and Steinmann P. "The discrete null space method for the energy-consistent integration of constrained mechanical systems. Part III: Flexible multibody dynamics". In: *Multibody Syst Dyn.* 19(1-2): 45–72, 2008. DOI: 10.1007/s11044-007-9056-4.

[93] Marino E. "Locking-free isogeometric collocation formulation for three-dimensional geometrically exact shear-deformable beams with arbitrary initial curvature". In: *Comput Method Appl M.* 324: 546–572, 2017. DOI: 10.1016/j.cma.2017.06.031.

[94] Marino E, Kiendl J, and De Lorenzis L. "Explicit isogeometric collocation for the dynamics of three-dimensional beams undergoing finite motions". In: *Comput Method Appl M.* 343: 530–549, 2019. DOI: 10.1016/j.cma.2018.09.005.

[95] Marsden JE and Ratiu TS. *Introduction to mechanics and symmetry.* Springer-Verlag, 1999.

[96] Marthinsen A. "Interpolation in Lie Groups". In: *SIAM J Numer Aanl.* 37(1): 269–285, 1999. DOI: 10.1137/s0036142998338861.

[97] Meier C, Popp A, and Wall WA. "An objective 3D large deformation finite element formulation for geometrically exact curved Kirchhoff rods". In: *Comput Method App M.* 278: 445–478, 2014. DOI: 10.1016/j.cma.2014.05.017.

[98] Meier C, Popp A, and Wall WA. "A locking-free finite element formulation and reduced models for geometrically exact Kirchhoff rods". In: *Comput Method App M.* 290: 314–341, 2015. DOI: 10.1016/j.cma.2015.02.029.

[99] Meier C, Popp A, and Wall WA. "Geometrically Exact Finite Element Formulations for Slender Beams: Kirchhoff–Love Theory Versus Simo–Reissner Theory". In: *Arch Comput Methods Eng.* 26(1): 163–243, 2019. DOI: 10.1007/s11831-017-9232-5.

[100] Melenk JM. "On condition numbers in hp-FEM with Gauss–Lobatto-based shape functions". In: *Journal of Computational and Applied Mathematics.* 139(1): 21–48, 2002. DOI: 10.1016/s0377-0427(01)00391-0.

[101] Milnor J. "Analytic Proofs of the "Hairy Ball Theorem" and the Brouwer Fixed Point Theorem". In: *Am Math Mon.* 85(7): 521–524, 1978. DOI: 10.2307/2320860.

[102] Müller A and Bischoff M. "A Consistent Finite Element Formulation of the Geometrically Non-linear Reissner-Mindlin Shell Model". In: *Arch Comput Method E.* 29(5): 3387–3434, 2022. DOI: 10.1007/s11831-021-09702-7.

[103] Muñoz JJ. "Modelling unilateral frictionless contact using the null-space method and cubic B-Spline interpolation". In: *Comput Methods Appl Mech Eng.* 197(9-12): 979–993, 2008. DOI: 10.1016/j.cma.2007.09.022.

[104] Muñoz JJ and Jelenić G. "Sliding contact conditions using the master-slave approach with application on geometrically non-linear beams". In: *Int J Solids Struct.* 41(24-25): 6963–6992, 2004. DOI: 10.1016/j.ijsolstr.2004.05.032.

[105] Muñoz JJ and Jelenić G. "Sliding contact conditions using the master-slave approach with application on geometrically non-linear beams". In: *Int J Solids Struct.* 41(24-25): 6963–6992, 2004. DOI: 10.1016/j.ijsolstr.2004.05.032.

[106] Muñoz JJ, Jelenić G, and Crisfield MA. "Master-slave approach for the modelling of joints with dependent degrees of freedom in flexible mechanisms". In: *Commun Numer Methods En.* 19(9): 689–702, 2003. DOI: 10.1002/cnm.627.

[107] Muñoz JJ and Jelenić G. "Sliding joints in 3D beams: Conserving algorithms using the master-slave approach". In: *Multibody Syst Dyn.* 16(3): 237–261, 2006. DOI: 10.1007/s11044-006-9025-3.

[108] Neff P, Lankeit J, and Madeo A. "On Grioli's minimum property and its relation to Cauchy's polar decomposition". In: *Int J Eng Sci.* 80: 209–217, 2014. DOI: 10.1016/j.ijengsci.2014.02.026.

[109] Nguyen VP, Anitescu C, Bordas SP, and Rabczuk T. "Isogeometric analysis: an overview and computer implementation aspects". In: *Math Comput Simulat.* 117: 89–116, 2015. DOI: 10.1016/j.matcom.2015.05.008.

[110] Piegl L and Tiller W. *The NURBS Book.* Berlin: Springer-Verlag Berlin Heidelberg GmbH, 1997.

[111] Reali A. "An isogeometric analysis approach for the study of structural vibrations". In: *J Earthq Eng.* 10(spec01): 1–30, 2006. DOI: 10.1080/13632460609350626.

[112] Reissner E. "On one-dimensional finite-strain beam theory: The plane problem". In: *ZAMP - Z Angew Math Phys.* 23(5): 795–804, 1972. DOI: 10.1007/bf01602645.

[113] Reissner E. "On one-dimensional large-displacement finite-strain beam theory". In: *Stud Appl Math.* 52(2): 87–95, 1973. DOI: 10.1002/sapm197352287.

[114] Romero I. "The interpolation of rotations and its application to finite element models of geometrically exact rods". In: *Computat Mech.* 34: 121–133, 2004. DOI: 10.1007/s00466-004-0559-z.

[115] Romero I. “A comparison of finite elements for nonlinear beams: the absolute nodal coordinate and geometrically exact formulations”. In: *Multibody Syst Dyn.* 20(1): 51–68, 2008. DOI: 10.1007/s11044-008-9105-7.

[116] Romero I. “An analysis of the stress formula for energy-momentum methods in nonlinear elastodynamics”. In: *Computat Mech.* 50: 603–610, 2012. DOI: 10.1007/s00466-012-0693-y.

[117] Romero I and Armero F. “An objective finite element approximation of the kinematics of geometrically exact rods and its use in the formulation of an energy–momentum conserving scheme in dynamics”. In: *Int J Numer Meth Eng.* 54(12): 1683–1716, 2002. DOI: 10.1002/nme.486.

[118] Romero I and Gebhardt CG. “ Variational principles for nonlinear Kirchhoff rods”. In: *Acta Mech.* 231: 625–647, 2020. DOI: 10.1007/s00707-019-02562-0.

[119] Rong J, Wu Z, Liu C, and Brüls O. “Geometrically exact thin-walled beam including warping formulated on the special Euclidean group SE (3)”. In: *Comput Method App M.* 369: 113062, 2020. DOI: 10.1016/j.cma.2020.113062.

[120] Sander O. “Geodesic finite elements for Cosserat rods”. In: *Int J Numer Meth Eng.* 82(13): 1645–1670, 2010. DOI: 10.1002/nme.2814.

[121] Sander O. “Geodesic finite elements on simplicial grids”. In: *Int J Numer Meth Eng.* 92(12): 999–1025, 2012. DOI: 10.1002/nme.4366.

[122] Sander O. “Geodesic finite elements of higher order”. In: *IMA J Numer Anal.* 36(1): 238–266, 2016. DOI: 10.1093/imanum/drv016.

[123] Sander O, Neff P, and Birsan M. “Numerical treatment of a geometrically nonlinear planar Cosserat shell model”. In: *Computat Mech.* 57: 817–841, 2016. DOI: 10.1007/s00466-016-1263-5.

[124] Sänger N. “Elemente für die Dynamik flexibler Mehrkörpersysteme”. PhD thesis. Universität Siegen, 2011. URL: https://dspace.ub.uni-siegen.de/handle/ubsi/561.

[125] Shoemake K. “Animating rotation with quaternion curves”. In: *Proceedings of the 12th annual conference on Computer graphics and interactive techniques.* 1985, pp. 245–254. DOI: 10.1145/325165.325242.

[126] Simo JC. “A finite strain beam formulation. The three-dimensional dynamic problem. Part I”. In: *Comput Methods Appl Mech Eng.* 49(1): 55–70, 1985. DOI: 10.1016/0045-7825(85)90050-7.

[127] Simo JC and Vu-Quoc L. “A Three-Dimensional Finite-Strain Rod Model. Part II: Computational Aspects”. In: *Comput Methods Appl Mech Eng.* 58: 79–116, 1986. DOI: 10.1016/0045-7825(86)90079-4.

[128] Simo JC and Vu-Quoc L. “On the dynamics of flexible beams under large overall motions-the plane case: Part I”. In: *J Appl Mech.* 53(4): 849–854, 1986. doi:10.1115/1.3171871.

[129] Simo JC and Vu-Quoc L. "On the dynamics of flexible beams under large overall motions-the plane case: Part II". In: *J Appl Mech.* 53(1): 855–863, 1986. doi:10.1016/0045-7825(86)90079-4.

[130] Simo JC and Tarnow N. "The discrete energy-momentum method. Conserving algorithms for nonlinear elastodynamics". In: *ZAMP - Z Angew Math Phys.* 43(5): 757–792, 1992. DOI: 10.1007/BF00913408.

[131] Simo JC, Tarnow N, and Doblare M. "Non-linear dynamics of three-dimensional rods: Exact energy and momentum conserving algorithms". In: *Int J Numer Meth Eng.* 38(9): 1431–1473, 1995. DOI: 10.1002/nme.1620380903.

[132] Sola J, Deray J, and Atchuthan D. *A micro Lie theory for state estimation in robotics.* Tech. rep. Technical Report IRI-TR-18-01. Institut de Ròbotica i Informàtica Industrial,Universitat Politècnica de Catalunya, 2018. URL: https://www.iri.upc.edu/files/scidoc/2089-A-micro-Lie-theory-for-state-estimation-in-robotics.pdf.

[133] Solin P, Segeth K, and Dolezel I. *Higher-order finite element methods.* Chapman & Hall/CRC, 2003.

[134] Sonneville V and Brüls O. "A few good reasons to consider a beam finite element formulation on the lie group SE (3)". In: *Proceedings of the ASME 2013 International Design Engineering Technical Conferences and Computers and Information in Engineering Conference.* ASME. Aug. 2013, pp. 1–12. DOI: 10.1115/detc2013-13099.

[135] Sonneville V, Brüls O, and Bauchau O. "Interpolation schemes for geometrically exact beams: a motion approach". In: *Int J Numer Meth Eng.* 112(9): 1129–1153, 2017. DOI: 10.1002/nme.5548.

[136] Sonneville V, Cardona A, and Brüls O. "Geometrically exact beam finite element formulated on the special Euclidean group SE (3)". In: *Comput Method App M.* 268: 451–474, 2014. DOI: 10.1016/j.cma.2013.10.008.

[137] Sprecher M. "Numerical methods for optimization and variational problems with manifold-valued data". PhD thesis. ETH Zurich, 2016.

[138] Spurrier RA. "Comment on "Singularity-Free Extraction of a Quaternion from a Direction-Cosine Matrix"". In: *J Spacecraft Rockets.* 15(4): 255–255, 1978. DOI: 10.2514/3.57311.

[139] Steinbrecher I, Mayr M, Grill MJ, Kremheller J, Meier C, and Popp A. "A mortar-type finite element approach for embedding 1D beams into 3D solid volumes". In: *Computat Mech.* 66: 1377–1398, 2020. DOI: 10.1007/s00466-020-01907-0.

[140] Steinbrecher I, Popp A, and Meier C. "Consistent coupling of positions and rotations for embedding 1D Cosserat beams into 3D solid volumes". In: *Computat Mech.* 69(3): 701–732, 2022. DOI: 10.1007/s00466-021-02111-4.

[141] Stillwell J. *Naive lie theory.* Springer Science & Business Media, 2008.

[142] Stuelpnagel J. "On the parametrization of the three-dimensional rotation group". In: *SIAM Rev.* 6(4): 422–430, 1964. DOI: 10.1137/1006093.

[143] Tasora A, Benatti S, Mangoni D, and Garziera R. “A geometrically exact isogeometric beam for large displacements and contacts”. In: *Comput Method App M.* 358: 112635, 2020. DOI: 10.1016/j.cma.2019.112635.

[144] Till J, Aloi V, and Rucker C. “Real-time dynamics of soft and continuum robots based on Cosserat rod models”. In: *Int J Robot Res.* 38(6): 723–746, 2019. DOI: 10.1177/0278364919842269.

[145] Tomec J and Jelenić G. “Momentum and near-energy conserving/decaying time integrator for beams with higher-order interpolation on SE (3)”. In: *Comput Method App M.* 419: 116665, 2024. DOI: 10.1016/j.cma.2023.116665.

[146] Vardoulakis I. *Cosserat continuum mechanics With Applications to Granular Media.* Vol. 87. Lecture Notes in Applied and Computational Mechanics. Springer, 2019.

[147] Vinberg EB. *A course in algebra.* 56. American Mathematical Soc., 2003.

[148] Wasmer P and Betsch P. “A projection-based quaternion discretization of the geometrically exact beam model”. In: *Int J Numer Meth Eng.* e7538, DOI: 10.1002/nme.7538.

[149] Weeger O, Yeung SK, and Dunn ML. “Isogeometric collocation methods for Cosserat rods and rod structures”. In: *Comput Method App M.* 316: 100–122, 2017. DOI: 10.1016/j.cma.2016.05.009.

[150] Willberg C, Duczek S, Perez JV, Schmicker D, and Gabbert U. “Comparison of different higher order finite element schemes for the simulation of Lamb waves”. In: *Comput Methods Appl Mech Eng.* 241: 246–261, 2012. DOI: 10.1016/j.cma.2012.06.011.

[151] Zupan E, Saje M, and Zupan D. “The quaternion-based three-dimensional beam theory”. In: *Comput Method App M.* 198(49-52): 3944–3956, 2009. DOI: 10.1016/j.cma.2009.09.002.

[152] Zupan E and Zupan D. “On conservation of energy and kinematic compatibility in dynamics of nonlinear velocity-based three-dimensional beams”. In: *Nonlinear Dynam.* 95(2): 1379–1394, 2019. DOI: 10.1007/s11071-018-4634-y.

Schriftenreihe des Instituts für Mechanik
ISSN 2363-4936

Hrsg. von: Prof. Dr.-Ing. habil. Peter Betsch, Prof. Dr.-Ing. habil. Thomas Seelig

Band 1 Marlon Franke
Discretisation techniques for large deformation computational contact elastodynamics.
ISBN 978-3-7315-0278-4

Band 2 Philipp Hempel
Constitutive modeling of amorphous thermoplastic polymers with special emphasis on manufacturing processes.
ISBN 978-3-7315-0550-1

Band 3 Martin Helbig
Mehrskalenmodellierung von Schädigung in gummimodifizierten thermoplastischen Kunststoffen.
ISBN 978-3-7315-0571-6

Band 4 Maik Dittmann
Isogeometric analysis and hierarchical refinement for multi-field contact problems.
ISBN 978-3-7315-0616-4

Band 5 Yinping Yang
Numerical methods for the inverse dynamics simulation of underactuated mechanical systems.
ISBN 978-3-7315-0626-3

Band 6 Alexander Janz
Structure-preserving space-time discretization in a mixed framework for multi-field problems in large strain elasticity.
ISBN 978-3-7315-0926-4

Band 7 Mark Georg Schiebl
Thermodynamically consistent space-time discretization of non-isothermal mechanical systems in the framework of GENERIC.
ISBN 978-3-7315-1117-5

Band 8 Jonas Hund
Characterisation and modelling of PC/ABS blends.
ISBN 978-3-7315-1157-1

Band 9 Jonathan Schulte
Multi-field modeling and simulation of fiber-reinforced polymers.
ISBN 978-3-7315-1251-6

Band 10 Robin Pfefferkorn
EAS Elements for Solid Mechanics – Mesh Distortion Insensitive and Hourglassing-Free Formulations with Increased Robustness.
ISBN 978-3-7315-1271-4

Band 11 Timo Ströhle
Inverse dynamics of underactuated flexible mechanical systems governed by quasi-linear hyperbolic partial differential equations.
ISBN 978-3-7315-1336-0

Band 12 Tobias Laschütza
Multiscale modelling of crazing in glassy polymers under cyclic loading.
ISBN 978-3-7315-1409-1

Band 13 Paul Wasmer
On the treatment of finite rotations in the discretization of a geometrically exact beam model.
ISBN 978-3-7315-1410-7